Assessment Item Listing for Weather and Climate

HOLT, RINEHART AND WINSTON

A Harcourt Classroom Education Company

Austin • New York • Orlando • Atlanta • San Francisco • Boston • Dallas • Toronto • London

Copyright © by Holt, Rinehart and Winston

All rights reserved. No part of this publication may be reproduced or transmitted in any form or by any means, electronic or mechanical, including photocopy, recording, or any information storage and retrieval system, without permission in writing from the publisher.

Teachers using HOLT SCIENCE AND TECHNOLOGY may photocopy complete pages in sufficient quantities for classroom use only and not for resale.

Art and Photo Credits

All work, unless otherwise noted, contributed by Holt, Rinehart and Winston.

Front cover: William Lesch/The Image Bank; (owl on cover, title page) Kim Taylor/Bruce Coleman, Inc.

Printed in the United States of America

ISBN 0-03-065518-8

2 3 4 5 6 082 05 04 03 02

CONTENTS

Introduction v
Installation and Startup vi
Getting Started viii
1 The Atmosphere 1
2 Understanding Weather 25
3 Climate 50

Introduction

The *Holt Science and Technology* Test Generator and *Assessment Item Listing*
The *Holt Science and Technology* Test Generator consists of a comprehensive bank of test items and the ExamView® Pro 3.0 software, which enables you to produce your own tests based on the items in the Test Generator and items you create yourself. Both Macintosh® and Windows® versions of the Test Generator are included on the *Holt Science and Technology* One-Stop Planner with Test Generator. Directions on pp. vi–vii of this book explain how to install the program on your computer. This *Assessment Item Listing* is a printout of all the test items in the *Holt Science and Technology* Test Generator.

ExamView Software
ExamView enables you to quickly create printed and on-line tests. You can enter your own questions in a variety of formats, including true/false, multiple choice, completion, problem, short answer, and essay. The program also allows you to customize the content and appearance of the tests you create.

Test Items
The *Holt Science and Technology* Test Generator contains a file of test items for each chapter of the textbook. Each item is correlated to the chapter objectives in the textbook and by difficulty level.

Item Codes
As you browse through this *Assessment Item Listing*, you will see that all test items of the same type appear under an identifying head. Each item is coded to assist you with item selection. Following is an explanation of the codes.

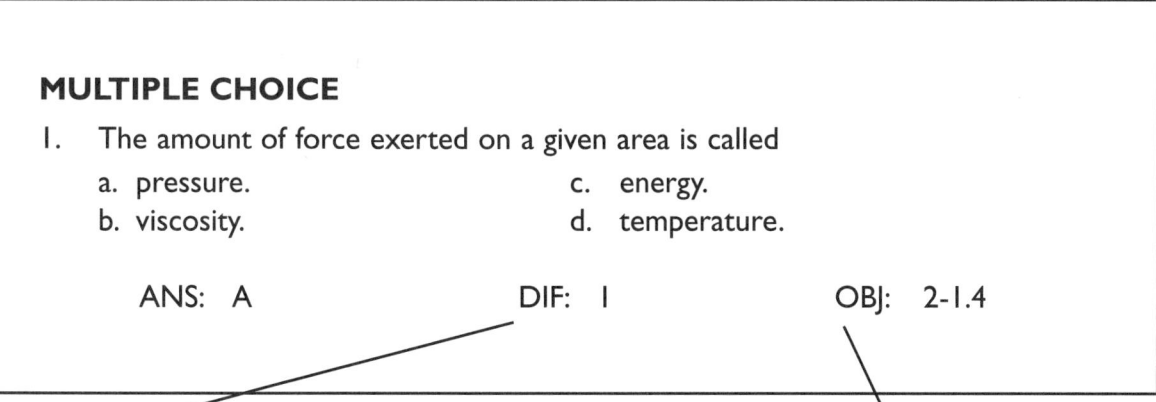

DIF defines the difficulty of the item.
  I requires recall of information.
  II requires analysis and interpretation of known information.
  III requires application of knowledge to new situations.

OBJ lists the chapter number, section number, and objective.
(2-1.4 = Chapter 2, Section 1, Objective 4)

INSTALLATION AND STARTUP

The Test Generator is provided on the One-Stop Planner. The Test Generator includes ExamView and all of the questions for the corresponding textbook. ExamView includes three components: Test Builder, Question Bank Editor, and Test Player. The Test Builder includes options to create, edit, print, and save tests. The Question Bank Editor lets you create or edit question banks. The Test Player is a separate program that your students can use to take on-line* (computerized or LAN-based) tests. Please refer to the ExamView User's Guide on the One-Stop Planner for complete instructions.

Before you can use the Test Generator, you must install ExamView and the test banks on your hard drive. The system requirements, installation instructions, and startup procedures are provided below.

SYSTEM REQUIREMENTS

To use ExamView, your computer must meet or exceed the following hardware requirements:

Windows®
- Pentium processor
- Windows 95®, Windows 98®, Windows 2000® (or a more recent version)
- color monitor (VGA-compatible)
- CD-ROM and/or high-density floppy disk drive
- hard drive with at least 7 MB space available
- 8 MB available memory (16 MB memory recommended)
- an Internet connection (if you wish to access the Internet testing features)*

Macintosh®
- PowerPC processor, 100 MHz
- System 7.5 (or a more recent version)
- color monitor (VGA-compatible)
- CD-ROM and/or high-density floppy disk drive
- hard drive with at least 7 MB space available
- 8 MB available memory (16 MB memory recommended)
- an Internet connection with System 8.6 or a more recent version (if you wish to access the Internet testing features)*

* You can use the Test Player to host tests on your personal or school Web site or local area network (LAN) at no additional charge. The ExamView Web site's Internet test-hosting service must be purchased separately. Visit www.examview.com to learn more.

INSTALLATION

Instructions for installing ExamView from the CD-ROM:

Windows®
Step 1
Turn on your computer.
Step 2
Insert the One-Stop Planner into the CD-ROM drive.
Step 3
Click the Start button on the taskbar, and choose the Run option.
Step 4
In the Open box, type "d:\setup.exe" (substitute the letter for your drive if it is not d:) and click OK.
Step 5
Follow the prompts on the screen to complete the installation process.

Macintosh®
Step 1
Turn on your computer.
Step 2
Insert the One-Stop Planner into the CD-ROM drive. When the CD-ROM icon appears on the desktop, double-click the icon.
Step 3
Double-click the ExamView Pro Installer icon.
Step 4
Follow the prompts on the screen to complete the installation process.

Instructions for installing ExamView from the Main Menu of the One-Stop Planner (Macintosh® or Windows®):

Follow steps 1 and 2 from above.
Step 3
Double-click One-Stop.pdf. (If you do not have Adobe Acrobat® Reader installed on your computer, install it before proceeding by clicking Reader Installer.)
Step 4
To advance to the Main Menu, click anywhere on the title screen.
Step 5
Click the Test Generator button.
Step 6
Click the appropriate Install ExamView button.
Step 7
Follow the prompts on the screen to complete the installation process.

GETTING STARTED

After you complete the installation process, follow these instructions to start ExamView. See the ExamView User's Guide on the One-Stop Planner for further instructions on the program's options for creating a test and editing a question bank.

Startup Instructions

Step 1
Turn on the computer.
Step 2
Windows®: Click the Start button on the taskbar. Highlight the Programs menu, and locate the ExamView Pro Test Generator folder. Select the ExamView Pro option to start the software.
Macintosh®: Locate and open the ExamView Pro folder. Double-click the ExamView Pro icon.
Step 3
The first time you run the software, you will be prompted to enter your name, school/institution name, and city/state. You are now ready to begin using ExamView.
Step 4
Each time you start ExamView, the Startup menu appears. Choose one of the options shown.
Step 5
Use ExamView to create a test or edit questions in a question bank.

Technical Support

If you have any questions about the Test Generator, call the Holt, Rinehart and Winston technical support line at 1-800-323-9239, Monday through Friday, 7:00 A.M. to 6:00 P.M., Central Standard Time. You can contact the Technical Support Center on the Internet at http://www.hrwtechsupport.com or by e-mail at tsc@hrwtechsupport.com.

Short Course I Chapter 1—The Atmosphere

MULTIPLE CHOICE

1. What is the most abundant gas in the air that we breathe?
   a. oxygen
   b. nitrogen
   c. hydrogen
   d. carbon dioxide

   ANS: B   DIF: I   OBJ: 1-1.1

2. The major source of oxygen for the Earth's atmosphere is
   a. sea water.
   b. the sun.
   c. plants.
   d. animals.

   ANS: C   DIF: I   OBJ: 1-1.1

3. The bottom layer of the atmosphere, where almost all weather occurs is the
   a. stratosphere.
   b. troposphere.
   c. thermosphere.
   d. mesosphere.

   ANS: B   DIF: I   OBJ: 1-1.4

4. About ____ percent of the solar energy that reaches the outer atmosphere is absorbed at the Earth's surface.
   a. 20
   b. 30
   c. 50
   d. 70

   ANS: C   DIF: I   OBJ: 1-2.1

5. The ozone layer is located in the
   a. stratosphere.
   b. troposphere.
   c. thermosphere.
   d. mesosphere.

   ANS: A   DIF: I   OBJ: 1-1.4

6. How does most thermal energy in the atmosphere move?
   a. conduction
   b. convection
   c. advection
   d. radiation

   ANS: B   DIF: I   OBJ: 1-2.2

7. The balance between incoming radiation and outgoing heat energy is called
   a. convection.
   b. conduction.
   c. greenhouse effect.
   d. radiation balance.

   ANS: D   DIF: I   OBJ: 1-2.3

8. Most of the United States is located in which prevailing wind belt?
   a. westerlies
   b. northeast trade winds
   c. southeast trade winds
   d. doldrums

   ANS: A   DIF: I   OBJ: 1-2.3

9. Which of the following is NOT a primary pollutant?
   a. car exhaust
   b. acid precipitation
   c. smoke from a factory
   d. fumes from burning plastic

   ANS: D DIF: I OBJ: 1-4.2

10. The Clean Air Act
    a. controls the amount of air pollutants that can be released from most sources.
    b. requires cars to run on fuels other than gasoline.
    c. requires many industries to use scrubbers.
    d. (a) and (c) only

    ANS: D DIF: I OBJ: 1-4.4

11. Wind occurs because air tends to move from regions of higher to lower
    a. latitude.
    b. pressure.
    c. nitrogen levels.
    d. humidity.

    ANS: B DIF: I OBJ: 1-3.1

12. As ____ increases, air pressure decreases.
    a. altitude
    b. radiation
    c. water vapor
    d. pollution

    ANS: A DIF: I OBJ: 1-1.2

13. Most of the heat from equatorial regions is moved toward the poles by
    a. convection.
    b. radiation.
    c. conduction.
    d. precipitation.

    ANS: A DIF: I OBJ: 1-2.2

14. An increase in the greenhouse effect would cause
    a. acid precipitation.
    b. conduction.
    c. convection.
    d. global warming.

    ANS: D DIF: I OBJ: 1-2.3

15. ____ removes ash and other particles from smokestacks.
    a. A scrubber
    b. Ozone
    c. Convection
    d. Radiation

    ANS: A DIF: I OBJ: 1-4.4

16. As you move upward through the atmosphere, the temperature
    a. increases.
    b. decreases.
    c. stays the same.
    d. varies.

    ANS: D DIF: I OBJ: 1-1.3

Holt Science and Technology
Copyright © by Holt, Rinehart and Winston. All rights reserved.

17. The Earth's atmosphere is divided into four layers based on
    a. pressure changes.
    b. altitude.
    c. temperature changes.
    d. the oxygen levels present.

    ANS: C   DIF: I   OBJ: 1-1.4

18. Which atmospheric layer is the densest?
    a. stratosphere
    b. troposphere
    c. mesosphere
    d. thermosphere

    ANS: B   DIF: I   OBJ: 1-1.4

19. In the stratosphere, temperature ____ with increasing altitude.
    a. decreases
    b. stays the same
    c. fluctuates
    d. increases

    ANS: D   DIF: I   OBJ: 1-1.4

20. The oxygen in the air you breathe is made up of ____ atoms.
    a. one
    b. two
    c. three
    d. four

    ANS: B   DIF: I   OBJ: 1-1.1

21. The *Upper Atmosphere Research Satellite* has detected large ____ in the mesosphere.
    a. meteoroids
    b. pockets of ozone
    c. wind storms
    d. amounts of electrically charged particles

    ANS: C   DIF: I   OBJ: 1-1.4

22. A high temperature means that
    a. particles are compacted together.
    b. particles are moving very fast.
    c. particles are moving very slowly.
    d. particles cannot move.

    ANS: B   DIF: I   OBJ: 1-1.4

23. Aurora borealis and aurora australis occur in the
    a. troposphere.
    b. stratosphere.
    c. mesosphere.
    d. ionosphere.

    ANS: D   DIF: I   OBJ: 1-1.4

24. The ionosphere reflects certain
    a. X rays.
    b. gamma rays.
    c. radio waves.
    d. ultraviolet radiation.

    ANS: C   DIF: I   OBJ: 1-1.4

25. Near the Earth's surface, air is heated by
    a. conduction.
    b. solar winds.
    c. ozone.
    d. convection.

    ANS: A   DIF: I   OBJ: 1-2.2

26. Average annual ____ in the Northern Hemisphere have been higher in the 1990s than at any other time in the past 600 years.
    a. air pressures
    b. surface temperatures
    c. air temperatures
    d. jet stream numbers

    ANS: B   DIF: I   OBJ: 1-2.3

27. The ____ is the lowest layer of the atmosphere.
    a. troposphere
    b. stratosphere
    c. mesosphere
    d. ionosphere

    ANS: A   DIF: I   OBJ: 1-1.4

28. The ____ is the atmospheric layer immediately above the troposphere.
    a. mesosphere
    b. stratosphere
    c. thermosphere
    d. ionosphere

    ANS: B   DIF: I   OBJ: 1-1.4

29. The uppermost atmospheric layer is the
    a. troposphere.
    b. stratosphere.
    c. mesosphere.
    d. thermosphere.

    ANS: D   DIF: I   OBJ: 1-1.4

30. The ____ is the coldest layer of the atmosphere.
    a. thermosphere
    b. stratosphere
    c. mesosphere
    d. troposphere

    ANS: C   DIF: I   OBJ: 1-1.4

31. Wind is created by differences in
    a. air temperature.
    b. humidity.
    c. air pressure.
    d. altitude.

    ANS: C   DIF: I   OBJ: 1-3.1

32. Compared to the poles, the air at the equator is warmer and
    a. less dense, creating an area of high pressure as it rises.
    b. less dense, creating an area of low pressure as it rises.
    c. more dense, creating an area of low pressure as it rises.
    d. more dense, creating an area of high pressure as it rises.

    ANS: B   DIF: I   OBJ: 1-3.1

33. Winds generally move from the
    a. poles to the equator.
    b. equator to the poles.
    c. east to the west.
    d. west to the east.

    ANS: A   DIF: I   OBJ: 1-3.2

34. In both hemispheres, winds that blow from 30° latitude to the equator are called
    a. westerlies.
    b. trade winds.
    c. polar easterlies.
    d. jet streams.

    ANS: B   DIF: I   OBJ: 1-3.2

35. Wind belts between 30° and 60° latitude in both the Northern and Southern Hemispheres are called
    a. westerlies.
    b. polar easterlies.
    c. trade winds.
    d. jet streams.

    ANS: A   DIF: I   OBJ: 1-3.2

36. Narrow belts of high-speed winds that blow in the upper troposphere and lower stratosphere are known as
    a. trade winds.
    b. westerlies.
    c. polar easterlies.
    d. jet streams.

    ANS: D   DIF: I   OBJ: 1-3.2

37. Wind belts that extend from the poles to 60° latitude in both hemispheres are called
    a. westerlies.
    b. trade winds.
    c. polar easterlies.
    d. jet streams.

    ANS: C   DIF: I   OBJ: 1-3.2

38. Each hemisphere has ____ wind belts as a result of pressure differences.
    a. one
    b. two
    c. three
    d. four

    ANS: C   DIF: I   OBJ: 1-3.2

39. In the ____, there is very little wind because of the warm rising air.
    a. doldrums
    b. horse latitudes
    c. convection cells
    d. mesosphere

    ANS: A   DIF: I   OBJ: 1-3.2

40. An area of low pressure around the equator is called the
    a. convection cells.
    b. horse latitudes.
    c. doldrums.
    d. jet streams.

    ANS: B   DIF: I   OBJ: 1-3.2

Holt Science and Technology
Copyright © by Holt, Rinehart and Winston. All rights reserved.

41. At the poles, cold air
    a. rises.
    b. warms.
    c. freezes.
    d. sinks.

    ANS: D   DIF: I   OBJ: 1-3.2

42. Which of the following global winds does NOT follow regular paths around the Earth?
    a. westerlies
    b. trade winds
    c. polar easterlies
    d. jet streams

    ANS: D   DIF: I   OBJ: 1-3.2

43. Which of the following statements describing a sea breeze is true?
    a. Air over the water is cooler, which creates an area of high pressure.
    b. Air over the water is cooler, which creates an area of low pressure.
    c. Air over the water is warmer, which creates an area of low pressure.
    d. Air over the water is warmer, which creates an area of high pressure.

    ANS: A   DIF: I   OBJ: 1-3.3

44. Which of the following statements describing a land breeze is true?
    a. Air over the water is cooler, which creates an area of high pressure.
    b. Air over the water is cooler, which creates an area of low pressure.
    c. Air over the water is warmer, which creates an area of low pressure.
    d. Air over the water is warmer, which creates an area of high pressure.

    ANS: C   DIF: I   OBJ: 1-3.3

45. At night, cool air sinks into a valley from the mountain peaks, creating a
    a. valley breeze.
    b. mountain breeze.
    c. land breeze.
    d. jet stream.

    ANS: B   DIF: I   OBJ: 1-3.3

46. During the day, warm air from a valley moves upslope, creating a
    a. valley breeze.
    b. mountain breeze.
    c. land breeze.
    d. sea breeze.

    ANS: A   DIF: I   OBJ: 1-3.3

47. Which of the following produces the greatest amount of pollution?
    a. human-made primary pollutants
    b. human-made secondary pollutants
    c. naturally made primary pollutants
    d. None of the above

    ANS: C   DIF: I   OBJ: 1-4.1

48. Which of the following is NOT considered to be a pollutant?
    a. dust
    b. pollen
    c. sea salt
    d. None of the above

    ANS: D   DIF: I   OBJ: 1-4.1

49. Automobile exhaust reacts with air and sunlight to form
    a. smog.
    b. ozone.
    c. a primary pollutant.
    d. acid precipitation.

    ANS: B   DIF: I   OBJ: 1-4.2

50. Ozone reacts with automobile exhaust to form
    a. smog.
    b. a primary pollutant.
    c. smoke.
    d. ash.

    ANS: A   DIF: I   OBJ: 1-4.2

51. Seventy percent of the carbon monoxide in the United States is produced by
    a. power plants.
    b. volcanoes.
    c. factories.
    d. fuel-burning vehicles.

    ANS: D   DIF: I   OBJ: 1-4.2

52. Ninety-six percent of the sulfur oxides released into the atmosphere is produced by
    a. power plants.
    b. volcanoes.
    c. factories.
    d. fuel-burning vehicles.

    ANS: A   DIF: I   OBJ: 1-4.2

53. Scientists have discovered that some chemicals released into the atmosphere react with ozone in the ozone layer, breaking the ozone down into
    a. nitrogen.
    b. carbon dioxide.
    c. oxygen.
    d. carbon monoxide.

    ANS: C   DIF: I   OBJ: 1-4.2

54. Which of the following is NOT an effect of air pollution?
    a. cancer
    b. allergies
    c. increased colds
    d. None of the above

    ANS: D   DIF: I   OBJ: 1-4.3

55. The Earth receives energy from the sun in the form of
    a. ozone.
    b. radiation.
    c. nitrogen.
    d. carbon dioxide.

    ANS: B   DIF: I   OBJ: 1-2.1

56. When radiation reaches the Earth's atmosphere, about 25 percent of it is
    a. reflected by the Earth's surface.
    b. absorbed by the Earth's surface.
    c. scattered and reflected by clouds and air.
    d. absorbed by ozone, clouds, and air.

    ANS: C   DIF: I   OBJ: 1-2.1

57. When radiation reaches the Earth's atmosphere, about 5 percent of it is
    a. reflected by the Earth's surface.
    b. absorbed by the Earth's surface.
    c. reflected by clouds and air.
    d. absorbed by ozone, clouds, and air.

    ANS: A   DIF: I   OBJ: 1-2.1

58. When radiation reaches the Earth's atmosphere, about 20 percent of it is
    a. reflected by the Earth's surface.
    b. absorbed by the Earth's surface.
    c. reflected by clouds and air.
    d. absorbed by ozone, clouds, and air.

    ANS: D   DIF: I   OBJ: 1-2.1

59. Which of the following does NOT involve a transfer of energy?
    a. radiation
    b. conduction
    c. convection
    d. none of the above

    ANS: D   DIF: I   OBJ: 1-2.1

60. Water vapor in the atmosphere is a
    a. gas.
    b. solid.
    c. liquid.
    d. All of the above

    ANS: A   DIF: I   OBJ: 1-1.1

61. Which of the following substances allows solar energy to pass through it but traps heat?
    a. wood
    b. metal
    c. glass
    d. cement

    ANS: C   DIF: I   OBJ: 1-2.3

Examine the diagram of Earth's global winds and answer the questions that follow.

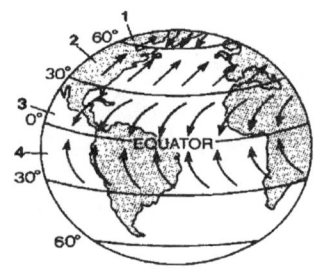

62. In which section do the westerlies occur?
    a. section 1
    b. section 2
    c. section 3
    d. section 4

    ANS: B   DIF: II   OBJ: 1-3.2

63. In which section do the southeast trade winds occur?
    a. section 1
    b. section 2
    c. section 3
    d. section 4

    ANS: D   DIF: II   OBJ: 1-3.2

64. In which section do the northeast trade winds occur?
    a. section 1
    b. section 2
    c. section 3
    d. section 4

    ANS: C   DIF: II   OBJ: 1-3.2

65. In which section do the polar easterlies occur?
    a. section 1
    b. section 2
    c. section 3
    d. section 4

    ANS: A   DIF: II   OBJ: 1-3.2

66. Where are the horse latitudes?
    a. at 0° latitude
    b. at 30° latitude
    c. at 60° latitude
    d. at the poles

    ANS: B   DIF: II   OBJ: 1-3.2

67. Where are the doldrums?
    a. at 0° latitude
    b. at 30° latitude
    c. at 60° latitude
    d. at the poles

    ANS: A   DIF: II   OBJ: 1-3.2

68. The uneven heating of the Earth produces pressure belts about every ____ on the Earth.
    a. 30° of longitude
    b. 60° of longitude
    c. 30° of latitude
    d. 60° of latitude

    ANS: C   DIF: I   OBJ: 1-3.2

COMPLETION

1. When you pick up a hot cup, heat is transferred from the cup to your hand primarily by _____. (conduction or convection)

   ANS: conduction   DIF: I   OBJ: 1-2.2

2. Acids formed in the air from sulfur compounds are an example of _____ pollutants. (primary or secondary)

   ANS: secondary   DIF: I   OBJ: 1-4.1

3. Winds that flow toward the poles in the opposite direction of the trade winds are called _____. (polar easterlies or westerlies)

   ANS: westerlies   DIF: I   OBJ: 1-3.2

4. The _____ is caused by gases in the atmosphere that absorb radiation and transfer heat. (greenhouse effect or Coriolis effect)

   ANS: greenhouse effect　　　　DIF: I　　　　OBJ: 1-2.3

5. Winds in the Northern Hemisphere curve to the right, and winds in the Southern Hemisphere curve to the left. This is known as the _____. (jet stream or Coriolis effect)

   ANS: Coriolis effect　　　　DIF: I　　　　OBJ: 1-3.2

6. _____ meet in an area of low pressure called the doldrums. (Trade winds or Jet streams)

   ANS: Trade winds　　　　DIF: I　　　　OBJ: 1-3.2

7. The _____ is a mixture of gases that surrounds the Earth.

   ANS: atmosphere　　　　DIF: I　　　　OBJ: 1-1.1

8. _____ is the height of an object above the Earth's surface.

   ANS: Altitude　　　　DIF: I　　　　OBJ: 1-1.2

9. _____ is the measure of the force with which air molecules push on a surface.

   ANS: Air pressure　　　　DIF: I　　　　OBJ: 1-1.2

10. _____ is a molecule that is made up of three oxygen atoms.

    ANS: Ozone　　　　DIF: I　　　　OBJ: 1-1.4

11. The layer of the stratosphere that absorbs solar energy in the form of ultraviolet radiation is called the _____ layer.

    ANS: ozone　　　　DIF: I　　　　OBJ: 1-1.4

12. The upper part of the mesosphere and the lower thermosphere is called the _____.

    ANS: ionosphere　　　　DIF: I　　　　OBJ: 1-1.4

13. _____ is a measure of the average energy of particles in motion.

    ANS: Temperature　　　　DIF: I　　　　OBJ: 1-1.4

14. _____ is the transfer of energy between objects at different temperatures.

    ANS: Heat　　　　DIF: I　　　　OBJ: 1-1.4

Holt Science and Technology
Copyright © by Holt, Rinehart and Winston. All rights reserved.

15. Electrically charged particles are called _____.

    ANS: ions   DIF: I   OBJ: 1-1.4

16. _____ is the transfer of energy as electromagnetic waves.

    ANS: Radiation   DIF: I   OBJ: 1-2.2

17. _____ is the transfer of thermal energy by the circulation or movement of a liquid or gas.

    ANS: Convection   DIF: I   OBJ: 1-2.2

18. A circular movement of air is called a _____.

    ANS: convection current   DIF: I   OBJ: 1-2.2

19. Carbon dioxide, a gas that traps thermal energy, is called a _____ gas.

    ANS: greenhouse   DIF: I   OBJ: 1-2.3

20. The circular patterns caused by the rising and sinking of air are called _____.

    ANS: convection cells DIF: I   OBJ: 1-3.2

21. The curving of moving objects by the Earth's rotation is called the _____.

    ANS: Coriolis effect   DIF: I   OBJ: 1-3.2

22. _____ winds are part of a pattern of air circulation that moves across the Earth.

    ANS: Global   DIF: I   OBJ: 1-3.2

23. _____ winds generally move short distances and can blow from any direction.

    ANS: Local   DIF: I   OBJ: 1-3.3

24. The _____ are the areas at about 30° north and 30° south latitude where sinking air creates an area of high pressure.

    ANS: horse latitudes   DIF: I   OBJ: 1-3.2

25. _____ pollutants are pollutants that are put directly into the air by human or natural activity.

    ANS: Primary   DIF: I   OBJ: 1-4.1

26. _____ pollutants are pollutants that form from chemical reactions that occur when primary pollutants come in contact with other primary pollutants or with naturally occurring substances.

    ANS: Secondary   DIF: I   OBJ: 1-4.1

27. Precipitation that contains acids from air pollution is called _____ precipitation.

    ANS: acid   DIF: I   OBJ: 1-4.1

28. _____ are chemical compounds that contain oxygen and other elements.

    ANS: Oxides   DIF: I   OBJ: 1-4.2

SHORT ANSWER

For each pair of terms, explain the difference in their meanings.

1. air pressure/altitude

    ANS:
    Air pressure is the measure of the force with which air molecules are pushing on a surface. Altitude is the height of an object above the Earth's surface.

    DIF: I   OBJ: 1-1.2

2. primary pollutant/secondary pollutant

    ANS:
    Primary pollutants are pollutants that are put directly into the air by human or natural activity. Secondary pollutants form from chemical reactions that occur when primary pollutants come in contact with other primary pollutants or with natural occurring substances.

    DIF: I   OBJ: 1-4.1

3. global wind/local wind

    ANS:
    Global winds are a part of air circulation that moves across the Earth. Local winds generally move short distances and can blow from any direction.

    DIF: I   OBJ: 1-3.3

4. troposphere/thermosphere

   ANS:
   The troposphere is the lowest layer of the Earth's atmosphere. The thermosphere is the uppermost atmospheric later.

   DIF: I   OBJ: 1-1.4

5. greenhouse effect/global warming

   ANS:
   The greenhouse effect is the Earth's natural heating process, by which gases in the atmosphere trap thermal energy. Global warming is a rise in average global temperatures possibly due to an increase in the greenhouse effect.

   DIF: I   OBJ: 1-2.3

6. convection/conduction

   ANS:
   Convection is the transfer of thermal energy by the circulation of a liquid or gas. Conduction is the transfer of energy from one material to another by direct contact.

   DIF: I   OBJ: 1-2.2

7. Explain why pressure decreases but temperature varies as altitude increases.

   ANS:
   Farther away from the Earth's surface, air pressure decreases because there are fewer gas molecules pushing down. Temperature varies due to the way solar energy is absorbed as it moves through the atmosphere.

   DIF: I   OBJ: 1-1.3

8. What causes air pressure?

   ANS:
   Air pressure is caused as gravity pulls molecules in the atmosphere toward the Earth.

   DIF: I   OBJ: 1-1.2

9. How can the thermosphere have high temperatures but not feel hot?

   ANS:
   In the thermosphere, particles are moving quickly, but because they are few and far apart, they cannot transfer much energy.

   DIF: I   OBJ: 1-1.3

10. Identify one characteristic of each layer of the atmosphere, and explain how that characteristic affects life on Earth.

    ANS:
    Sample answer: The gases in the troposphere make life on Earth possible. The stratosphere contains the ozone layer, which absorbs ultraviolet radiation. The mesosphere is the coldest layer of the atmosphere and it may affect weather patterns on Earth. The thermosphere contains the ionosphere, which absorbs harmful solar energy.

    DIF: II   OBJ: 1-1.4

11. Describe three things that can happen to radiation when it reaches the Earth's atmosphere.

    ANS:
    Answers will vary. Sample answer: Radiation can be absorbed by the Earth's surface. It can be absorbed by ozone, clouds, and the atmosphere or reflected by the Earth's surface and clouds.

    DIF: I   OBJ: 1-2.1

12. How is energy transferred through the atmosphere?

    ANS:
    Answers will vary. Sample answer: Energy is transferred through the atmosphere through radiation, conduction, and convection.

    DIF: I   OBJ: 1-2.2

13. What is the greenhouse effect?

    ANS:
    The greenhouse effect is the Earth's natural heating process by which gases in the atmosphere trap thermal energy.

    DIF: I   OBJ: 1-2.3

14. How does the process of convection rely on conduction?

    ANS:
    Answers will vary. Sample answer: The air directly above the Earth's surface is heated by conduction. This warm air is then circulated through the atmosphere by convection currents.

    DIF: II   OBJ: 1-2.2

15. How does the Coriolis effect affect wind movement?

    ANS:
    The Coriolis effect prevents winds from blowing directly north or south. Due to the Coriolis effect, trade winds in the Northern Hemisphere curve to the right, and trade winds in the Southern Hemisphere curve to the left.

    DIF: I   OBJ: 1-3.2

16. What causes winds?

    ANS:
    Winds are caused by the unequal heating of the Earth's surface and by pressure differences.

    DIF: I   OBJ: 1-3.1

17. Compare and contrast global winds and local winds.

    ANS:
    Local winds travel short distances and can blow from any direction. Global winds travel long distances and travel in specific directions.

    DIF: I   OBJ: 1-3.3

18. Suppose you are vacationing at the beach. It is daytime and you want to go swimming in the ocean. You know the beach is near your hotel, but you don't know what direction it is in. How might the local wind help you find the ocean?

    ANS:
    During the day, a sea breeze is caused by the cooler air over the water moving toward the land. Walking toward the sea breeze would lead you to the ocean.

    DIF: II   OBJ: 1-3.3

19. How can the air inside a building be more polluted than the air outside?

    ANS:
    Answers will vary. Indoor air is polluted by household cleaners, air fresheners, smoke from cooking, as well as industrial compounds found in carpets, paints, building materials, and furniture.

    DIF: I   OBJ: 1-4.2

20. Why might it be difficult to establish a direct link between air pollution and health problems?

    ANS:
    Answers will vary. Accept all reasonable responses.

    DIF: I   OBJ: 1-4.3

21. How has the Clean Air Act helped to reduce air pollution?

    ANS:
    The Clean Air Act gives the EPA the authority to control the amount of air pollutants that can be released from any source. The EPA also monitors air quality; if the air quality worsens, the EPA can set stricter standards.

    DIF: I   OBJ: 1-4.4

22. How is the water cycle affected by air pollution?

    ANS:
    Answers will vary. Sample answer: Rainwater can become more acidic as a result of air pollution.

    DIF: II   OBJ: 1-4.2

23. What are the two main gases in Earth's atmosphere?

    ANS:
    Nitrogen and oxygen are the two main gases in Earth's atmosphere.

    DIF: I   OBJ: 1-1.1

24. What is atmospheric pressure?

    ANS:
    Atmospheric pressure is the force exerted by molecules of air on a surface.

    DIF: I   OBJ: 1-1.2

25. Name the layers of the atmosphere, starting with the one closest to Earth.

    ANS:
    The layers of the atmosphere include the troposphere, stratosphere, mesosphere, and the thermosphere.

    DIF: I   OBJ: 1-1.4

26. What is the ozone layer, and why is it important to Earth?

    ANS:
    The ozone layer is a layer of ozone molecules in the stratosphere. The layer filters ultraviolet radiation from the sun and prevents much of this radiation from reaching Earth.

    DIF: I   OBJ: 1-1.4

27. Explain how density affects energy transfer in the air.

    ANS:
    The less dense the air is, the less effective it is at transferring heat. Particles that are farther apart, or less densely packed, are less likely to collide with other particles. Particles must collide with one another in order to transfer heat.

    DIF: I   OBJ: 1-2.2

28. What is radiation?

    ANS:
    Radiation is energy transferred as electromagnetic waves.

    DIF: I   OBJ: 1-2.1

29. A metal spoon left in a bowl of hot soup feels hot. Which process—radiation, conduction, or convection—is mainly responsible for heating the spoon?

    ANS:
    Conduction is mainly responsible for heating the spoon.

    DIF: I   OBJ: 1-2.2

30. What is a convection current?

    ANS:
    A convection current is the continual, circular movement of warm and cool particles in a liquid or gas.

    DIF: I   OBJ: 1-2.2

31. How does a greenhouse stay warm?

    ANS:
    Sunlight goes through the glass. Objects in the structure absorb some of the radiant energy. In turn, the objects radiate thermal energy. The glass prevents the energy from escaping, which warms the greenhouse.

    DIF: I   OBJ: 1-2.3

32. What is wind?

    ANS:
    Wind is air that flows between air masses of different pressures and temperatures.

    DIF: I   OBJ: 1-3.1

33. How does air temperature over landmasses and adjacent bodies of water change between day and night?

    ANS:
    During the day, the air is cooler over water. At night, the air is cooler over land.

    DIF: I   OBJ: 1-3.2

34. What is the Coriolis effect?

    ANS:
    The Coriolis effect is the deflection of moving objects due to Earth's rotation.

    DIF: I   OBJ: 1-3.2

35. Compare and contrast the trade winds and the westerlies in the Northern Hemisphere.

    ANS:
    Both are global wind systems that curve due to the Coriolis effect. Both result from differences in air pressure and temperature. The trade winds that lie between the equator and 30° north latitude blow from the northeast to the southwest. The westerlies lying between 30° and 60° north latitude blow from the southwest to the northeast.

    DIF: I   OBJ: 1-3.2

36. What are two kinds of breezes that result from local topography?

    ANS:
    Mountain and valley breezes result from local topography.

    DIF: I   OBJ: 1-3.3

37. Classify each of the following as either a primary or secondary air pollutant: smog, tobacco smoke, chalk dust, and acid rain.

    ANS:
    Tobacco smoke and chalk dust are primary pollutants, while smog and acid rain are secondary pollutants.

    DIF: I   OBJ: 1-4.1

38. What are the three sources of outdoor air pollution?

    ANS:
    Motor vehicles, industries, and electric power plants are sources of outdoor air pollution.

    DIF: I   OBJ: 1-4.2

39. What are two health problems that can result form breathing polluted air?

    ANS:
    Health problems that can result form breathing polluted air may include some or all of the following: dizziness, headaches, burning, itchy eyes, runny nose, coughing, shortness of breath, sore throat, lung cancer and other respiratory diseases, chest pain, colds and allergies.

    DIF: I   OBJ: 1-4.3

40. Why does the atmosphere become less dense as altitude increases?

    ANS:
    As altitude increases, there are fewer gas molecules. Gravity pulls much of the atmosphere's gas molecules close to the Earth's surface.

    DIF: I   OBJ: 1-1.2

41. Explain why air rises when it is heated.

    ANS:
    Air rises as it is heated because it becomes less dense.

    DIF: I   OBJ: 1-1.3

42. What causes temperature changes in the atmosphere?

    ANS:
    The temperature differences in the atmosphere result mainly from the way solar energy is absorbed as it moves downward through the atmosphere. Some layers are warmer because they contain gases that absorb solar energy.

    DIF: I   OBJ: 1-1.3

43. What are secondary pollutants, and how are they formed? Give an example.

    ANS:
    Secondary pollutants form when a primary pollutant reacts with other primary pollutants or with naturally occurring substances. Smog and ozone are examples of secondary pollutants.

    DIF: I   OBJ: 1-4.2

44. Use the following terms to create a concept map: *altitude, air pressure, temperature, atmosphere.*

    ANS:

    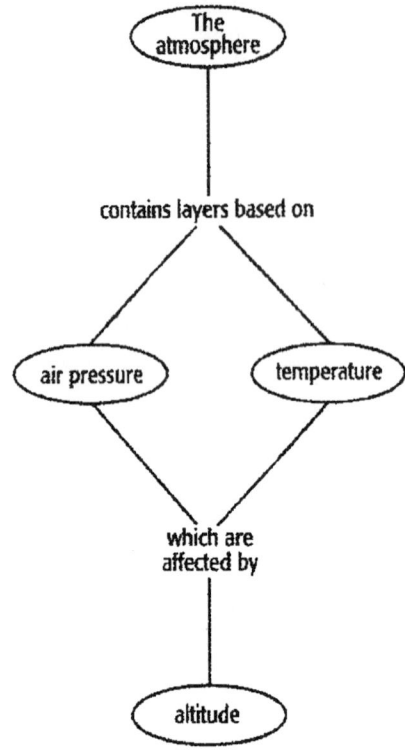

    DIF: II OBJ: 1-1.3

45. What is the relationship between the greenhouse effect and global warming?

    ANS:
    An increase in the greenhouse effect could result in global warming.

    DIF: II OBJ: 1-2.3

46. How do you think the Coriolis effect would change if the Earth were to rotate twice as fast? Explain.

    ANS:
    The Coriolis effect would be more pronounced if the Earth rotated twice as fast. Winds are affected by the rotation of the Earth; if the speed were increased, the curvature would be more pronounced.

    DIF: II OBJ: 1-3.2

47. Without the atmosphere, the Earth's surface would be very different. What are several ways that the atmosphere affects the Earth?

ANS:
Answers will vary. Sample answer: The atmosphere protects living organisms from harmful radiation from the sun. Without the atmosphere, more of this radiation would reach the Earth's surface.

DIF: II   OBJ: 1-1.4

48. Wind speed is measured in miles per hour and in knots. One mile (statute mile or land mile) is 5,280 ft. One nautical mile (or sea mile) is 6,076 ft. Speed in nautical miles is measured in knots. Calculate the wind speed in knots if the wind is blowing at 25 miles/h.

ANS:
22 knots

DIF: II   OBJ: 1-3.1

Use the wind-chill chart to answer the questions that follow.

Wind-Chill Chart

Wind Speed		Actual thermometer reading (°F)				
Knots	mph	40	30	20	10	0
		Equivalent temperature (°F)				
Calm		40	30	20	10	0
4	5	37	27	16	6	−5
9	10	28	16	4	−9	−21
13	15	22	9	−5	−18	−36
17	20	18	4	−10	−25	−39
22	25	16	0	−15	−29	−44
26	30	13	−2	−18	−33	−48
30	35	11	−4	−20	−35	−49

49. If the wind speed is 20 miles/h and the temperature is 40°F, how cold will the air seem?

ANS:
18°F

DIF: II   OBJ: 1-3.1

50. If the wind speed is 30 miles/h and the temperature is 20°F, how cold will the air seem?

    ANS:
    −18°F

    DIF: II   OBJ: 1-3.1

51. Explain why nighttime land breezes occur in areas close to the ocean.

    ANS:
    At night, the air over land gets cool more quickly than the air over water. The cool air over land creates an area of high pressure. The warmer air over water creates an area of low pressure. Because air always moves from areas of high pressure to areas of low pressure, air moves from the land to the ocean and causes a land breeze.

    DIF: I   OBJ: 1-3.3

52. Explain why mountain breezes occur.

    ANS:
    At night, mountains cool faster than the valley below. The cool air on the mountain is more dense than the warm air below. The cool, dense air moves down the mountain, creating a mountain breeze

    DIF: I   OBJ: 1-3.3

53. Explain how gases in the Earth's atmosphere can be compared to the glass covering a greenhouse.

    ANS:
    Both gases in the Earth's atmosphere and the glass covering a greenhouse allow solar energy to pass through them but stop heat energy from escaping to the outside.

    DIF: I   OBJ: 1-2.3

54. Suppose the stratosphere became covered in a thick blanket of volcanic dust. How would this affect the temperature of the air in the troposhpere?

    ANS:
    The temperature in the troposphere would drop. Normally, the sun's radiation reaches the troposphere and the energy from radiation is transferred into heat. If the stratosphere was covered in a layer of volcanic dust, less of the sun's radiation would reach the Earth's surface, and less heat would be produced.

    DIF: II   OBJ: 1-1.3

The graph below shows atmospheric carbon dioxide levels at a site in Hawaii from 1958 to 1988. Examine the graph and answer the questions that follow.

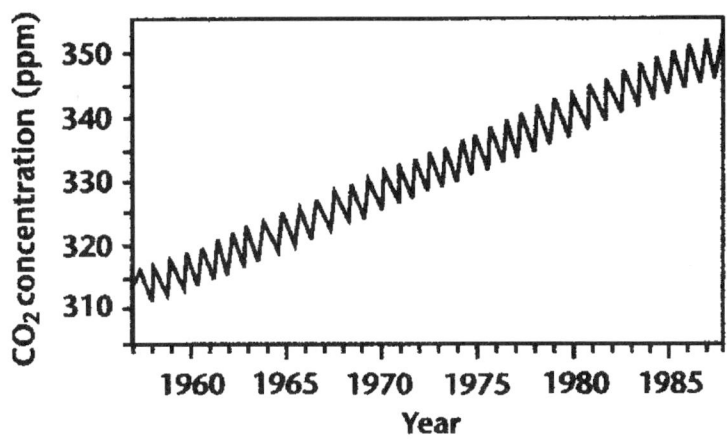

55. According to the graph, how have carbon dioxide levels changed from 1958 to 1988?

    ANS:
    Carbon dioxide levels have slowly increased from 1958 to 1988.

    DIF: II   OBJ: 1-4.2

56. Why might scientists be concerned about the trend shown in the graph?

    ANS:
    Sample answer: Some scientists believe that an increase in carbon dioxide levels has led to an increase in global temperatures. Higher global temperatures could cause serious climatic and weather changes.

    DIF: II   OBJ: 1-4.2

57. Use the following terms to complete the concept map below: *heat, density, thermosphere, ionosphere, particle movement, solar energy, temperature*

ANS:

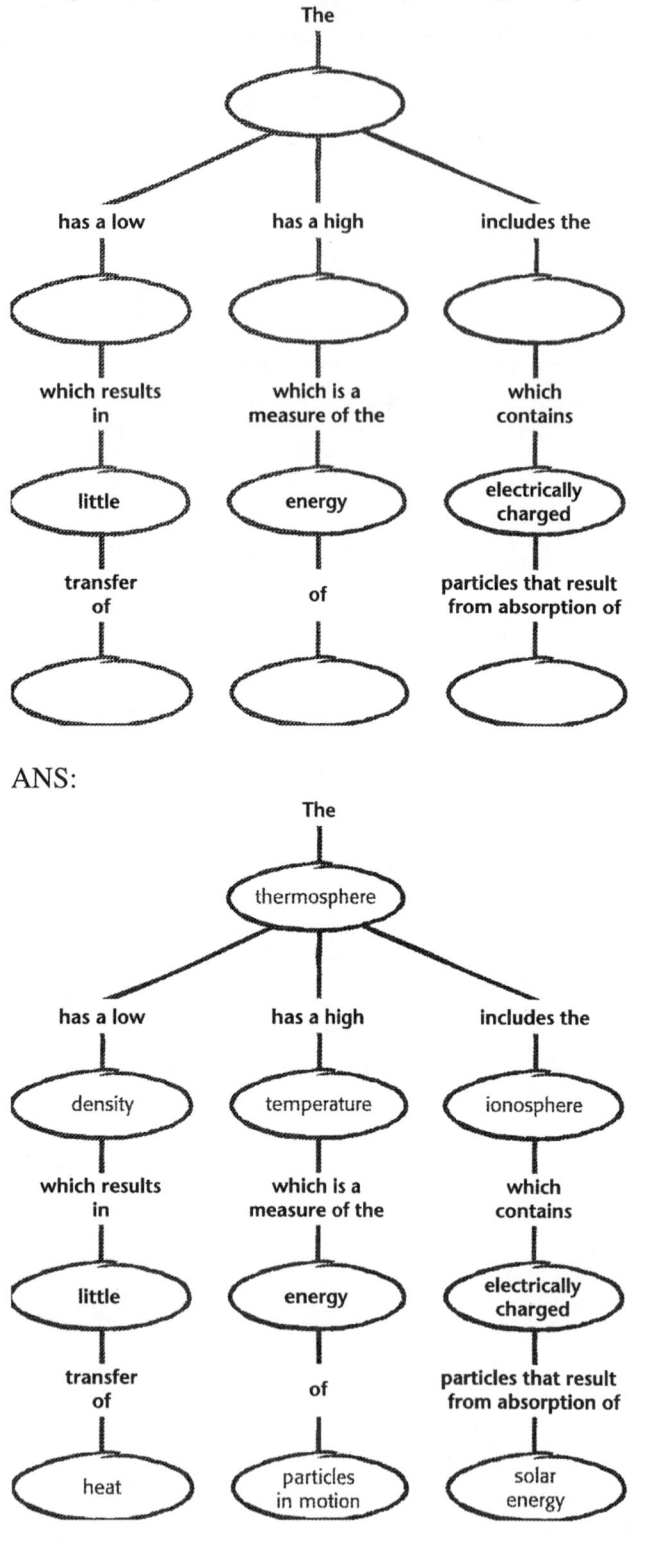

DIF: II   OBJ: 1-1.4

Short Course I Chapter 2—Understanding Weather

MULTIPLE CHOICE

1. The process of liquid water changing to gas is called
   a. precipitation.
   b. condensation.
   c. evaporation.
   d. water vapor.

   ANS: C   DIF: I   OBJ: 2-1.1

2. What is the relative humidity of air at its dew-point temperature?
   a. 0 percent
   b. 50 percent
   c. 75 percent
   d. 100 percent

   ANS: D   DIF: I   OBJ: 2-1.3

3. Which of the following is NOT a type of condensation?
   a. fog
   b. cloud
   c. snow
   d. dew

   ANS: C   DIF: I   OBJ: 2-1.3

4. High clouds made of ice crystals are called ____ clouds.
   a. stratus
   b. cumulus
   c. nimbostratus
   d. cirrus

   ANS: D   DIF: I   OBJ: 2-1.4

5. Large thunderhead clouds that produce precipitation are called ____ clouds.
   a. nimbostratus
   b. cumulonimbus
   c. cumulus
   d. stratus

   ANS: B   DIF: I   OBJ: 2-1.4

6. Strong updrafts within a thunderhead can produce
   a. snow.
   b. rain.
   c. sleet.
   d. hail.

   ANS: D   DIF: I   OBJ: 2-1.5

7. A maritime tropical air mass contains
   a. warm, wet air.
   b. cold, moist air.
   c. warm, dry air.
   d. cold, dry air.

   ANS: A   DIF: I   OBJ: 2-2.1

8. A front that forms when a warm air mass is trapped between cold air masses and forced to rise is called a(n)
   a. stationary front.
   b. warm front.
   c. occluded front.
   d. cold front.

   ANS: C   DIF: I   OBJ: 2-2.3

Holt Science and Technology
Copyright © by Holt, Rinehart and Winston. All rights reserved.

9. A severe storm that forms as a rapidly rotating funnel cloud is called a
   a. hurricane.
   b. tornado.
   c. typhoon.
   d. thunderstorm.

   ANS: B   DIF: I   OBJ: 2-3.3

10. The lines on a weather map connecting points of equal atmospheric pressure are called
    a. contour lines.
    b. highs.
    c. isobars.
    d. lows.

    ANS: C   DIF: I   OBJ: 2-4.2

11. To measure air pressure most accurately, you should use a mercury
    a. barometer.
    b. thermometer.
    c. psychrometer.
    d. wind vane.

    ANS: A   DIF: I   OBJ: 2-4.1

12. A windsock does NOT
    a. consist of a cone-shaped bag.
    b. measure wind speed.
    c. measure wind direction.
    d. allow wind to pass through it.

    ANS: B   DIF: I   OBJ: 2-4.1

13. Isobars indicate
    a. pressure.
    b. rainfall.
    c. snow.
    d. wind speed.

    ANS: A   DIF: I   OBJ: 2-4.2

14. Which of the following is NOT used to collect weather-related data from the upper atmosphere?
    a. weather balloon
    b. Doppler radar
    c. psychrometer
    d. orbital satellite

    ANS: C   DIF: I   OBJ: 2-4.1

15. Unstable atmospheric conditions lead to the formation of lightning and thunder from towering
    a. nimbostratus clouds.
    b. alto cumulus clouds.
    c. altostratus clouds.
    d. cumulonimbus clouds.

    ANS: D   DIF: I   OBJ: 2-3.2

16. Air's ability to hold water vapor increases as ____ increases.
    a. wind speed
    b. temperature
    c. air pressure
    d. All of the above

    ANS: B   DIF: II   OBJ: 2-1.3

Holt Science and Technology
Copyright © by Holt, Rinehart and Winston. All rights reserved.

17. Which of the following causes the most damage during a hurricane?
    a. water spouts
    b. high winds
    c. lightning
    d. storm surges

    ANS: D   DIF: I   OBJ: 2-3.3

18. Lightning is seen before thunder is heard because
    a. storm winds slow down sound waves.
    b. sound is created slowly.
    c. light travels faster than sound.
    d. ice crystals in clouds absorb sounds.

    ANS: C   DIF: I   OBJ: 2-3.1

Study the illustration below, and answer the questions that follow.

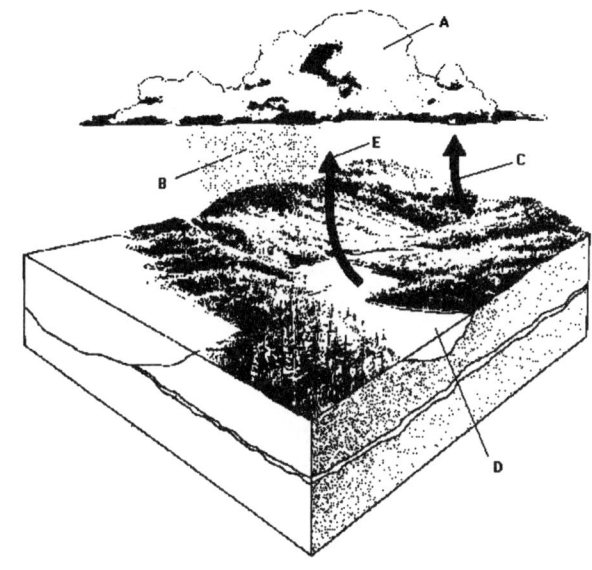

19. The illustration above is of the
    a. water cycle.
    b. carbon cycle.
    c. nitrogen cycle.
    d. greenhouse effect.

    ANS: A   DIF: II   OBJ: 2-1.1

20. What process occurs at **A**?
    a. precipitation
    b. condensation
    c. evaporation
    d. transpiration

    ANS: B   DIF: II   OBJ: 2-1.1

21. What process occurs at **B**?
    a. precipitation
    b. evaporation
    c. runoff
    d. transpiration

    ANS: A   DIF: II   OBJ: 2-1.1

22. What process occurs at **C**?
    a. condensation
    b. precipitation
    c. runoff
    d. transpiration

    ANS: D   DIF: II   OBJ: 2-1.1

23. What process occurs at **D**?
    a. transpiration
    b. precipitation
    c. runoff
    d. condensation

    ANS: C   DIF: II   OBJ: 2-1.1

24. What process occurs at **E**?
    a. runoff
    b. precipitation
    c. evaporation
    d. condensation

    ANS: C   DIF: II   OBJ: 2-1.1

25. Clouds are formed by
    a. evaporation
    b. precipitation
    c. transpiration
    d. condensation

    ANS: D   DIF: I   OBJ: 2-1.1

26. ____ occurs when rain, snow, sleet, or hail falls from the clouds onto the Earth's surface.
    a. Precipitation
    b. Condensation
    c. Evaporation
    d. Transpiration

    ANS: A   DIF: I   OBJ: 2-1.1

27. ____ occurs when water vapor cools and changes back into liquid droplets.
    a. Evaporation
    b. Condensation
    c. Transpiration
    d. Runoff

    ANS: B   DIF: I   OBJ: 2-1.1

28. ____ occurs when liquid water changes into water vapor.
    a. Condensation
    b. Transpiration
    c. Evaporation
    d. Precipitation

    ANS: C   DIF: I   OBJ: 2-1.1

29. ____ is the process by which plants release water vapor into the air through their leaves.
    a. Evaporation
    b. Transpiration
    c. Condensation
    d. Precipitation

    ANS: B   DIF: I   OBJ: 2-1.1

30. ____ is water, usually from precipitation, that flows across land and collects in rivers, streams, and eventually the ocean.
    a. Runoff
    b. Humidity
    c. Relative humidity
    d. Condensation

    ANS: A   DIF: I   OBJ: 2-1.1

31. ____ is the amount of water vapor or moisture in the air.
    a. Dew point
    b. Humidity
    c. Relative humidity
    d. Weather

    ANS: B   DIF: I   OBJ: 2-1.2

32. ____ is the amount of moisture the air contains compared with the maximum amount it can hold at a particular temperature.
    a. Weather
    b. Humidity
    c. Relative humidity
    d. Dew point

    ANS: C   DIF: I   OBJ: 2-1.2

33. Suppose that 1 m^3 of air at a certain temperature can hold 30 g of water vapor. However, you know that the air actually contains 15 g of water vapor. What is the relative humidity?
    a. 25 percent
    b. 50 percent
    c. 75 percent
    d. 100 percent

    ANS: B   DIF: II   OBJ: 2-1.2

34. Which device measures relative humidity?
    a. anemometer
    b. barometer
    c. psychrometer
    d. thermometer

    ANS: C   DIF: I   OBJ: 2-1.2

35. Suppose that 1 m^3 of air at a certain temperature can hold 20 g of water vapor. However, you know that the air actually contains 5 g of water vapor. What is the relative humidity?
    a. 25 percent
    b. 50 percent
    c. 75 percent
    d. 100 percent

    ANS: A   DIF: II   OBJ: 2-1.2

36. Before condensation can occur, what must the relative humidity be?
    a. 25 percent
    b. 50 percent
    c. 75 percent
    d. 100 percent

    ANS: D   DIF: I   OBJ: 2-1.3

37. Air can become saturated when water vapor is added to the air through
    a. evaporation.
    b. transpiration.
    c. precipitation
    d. Both (a) and (b)

    ANS: D   DIF: I   OBJ: 2-1.3

38. Which statement best describes condensation?
    a. It only occurs in cold areas.
    b. It occurs with no relative humidity.
    c. There must be a surface to condense on.
    d. It is the same as precipitation.

    ANS: C   DIF: I   OBJ: 2-1.3

39. Puffy, white clouds that tend to have flat bottoms are called
    a. stratus clouds.
    b. cirrus clouds.
    c. cumulus clouds.
    d. nimbus clouds.

    ANS: C   DIF: I   OBJ: 2-1.4

40. Clouds that form in layers are called
    a. stratus clouds.
    b. nimbus clouds.
    c. cumulus clouds.
    d. cirrus clouds.

    ANS: A   DIF: I   OBJ: 2-1.4

41. Thin, feathery, white clouds found at high altitudes are called
    a. stratus clouds.
    b. cumulus clouds.
    c. nimbus clouds.
    d. cirrus clouds.

    ANS: D   DIF: I   OBJ: 2-1.4

42. The most common form of precipitation is liquid water that falls from the clouds to Earth. This type of precipitation is called
    a. snow.
    b. rain.
    c. hail.
    d. sleet.

    ANS: B   DIF: I   OBJ: 2-1.5

43. ____, the most common form of solid precipitation, forms when temperatures are so cold that water vapor changes directly to a solid.
    a. Rain
    b. Hail
    c. Sleet
    d. Snow

    ANS: D   DIF: I   OBJ: 2-1.5

44. ____, also called freezing rain, forms when rain falls through a layer of freezing air.
    a. Sleet
    b. Hail
    c. Snow
    d. Flurries

    ANS: A   DIF: I   OBJ: 2-1.5

45. Solid precipitation that falls as balls or lumps of ice is called
    a. snow.
    b. hail.
    c. rain.
    d. snow.

    ANS: B   DIF: I   OBJ: 2-1.5

46. A _____ air mass is a wet air mass that forms over water.
    a. maritime (m)
    b. continental (c)
    c. polar (P)
    d. tropical (T)

    ANS: A　　　　DIF: I　　　　OBJ: 2-2.1

47. A _____ air mass is a dry air mass that forms over land.
    a. polar (P)
    b. continental (c)
    c. maritime (m)
    d. tropical (T)

    ANS: B　　　　DIF: I　　　　OBJ: 2-2.1

48. A _____ air mass is a cold air mass that forms over the polar regions.
    a. continental (c)
    b. tropical (T)
    c. maritime (m)
    d. polar (P)

    ANS: D　　　　DIF: I　　　　OBJ: 2-2.1

49. A _____ air mass is a warm air mass that develops over the Tropics.
    a. polar (P)
    b. tropical (T)
    c. maritime (m)
    d. continental (c)

    ANS: B　　　　DIF: I　　　　OBJ: 2-2.1

50. A _____ air mass forms over the North Pacific Ocean and affects the Pacific Coast.
    a. maritime polar (mP)
    b. maritime tropical (mT)
    c. continental polar (cP)
    d. continental tropical (cT)

    ANS: A　　　　DIF: II　　　　OBJ: 2-2.2

51. A _____ air mass forms over the North Atlantic Ocean and affects New England and the eastern part of Canada.
    a. maritime tropical (mT)
    b. maritime polar (mP)
    c. continental polar (cP)
    d. continental tropical (cT)

    ANS: B　　　　DIF: II　　　　OBJ: 2-2.2

52. A _____ air mass develops over warm areas in the Gulf of Mexico and the North Atlantic Ocean and move across the East Coast and into the Midwest.
    a. continental tropical (cT)
    b. maritime polar (mP)
    c. continental polar (cP)
    d. maritime tropical (mT)

    ANS: D　　　　DIF: II　　　　OBJ: 2-2.2

53. A _____ air mass forms over the deserts of northern Mexico and in the southwestern United States. It influences weather in the United States only during the summer as it moves northeastward, bringing clear, dry, and very hot weather.
    a. continental tropical (cT)
    b. continental polar (cP)
    c. maritime polar (mP)
    d. maritime tropical (mT)

    ANS: A　　　　DIF: II　　　　OBJ: 2-2.2

54. A(n) _____ occurs when a cold air mass meets and displaces a warm air mass.
    a. occluded front
    b. stationary front
    c. cold front
    d. warm front

    ANS: C   DIF: I   OBJ: 2-2.3

55. A(n) _____ occurs when a warm air mass meets and overrides a cold air mass.
    a. stationary front
    b. occluded front
    c. cold front
    d. warm front

    ANS: D   DIF: I   OBJ: 2-2.3

56. A(n) _____ occurs when a faster-moving cold air mass overtakes a slower-moving warm air mass and forces the warm air up.
    a. stationary front
    b. occluded front
    c. cold front
    d. warm front

    ANS: B   DIF: I   OBJ: 2-2.3

57. Cooler weather usually follows a(n) _____ front because the warm air is pushed away from the Earth's surface.
    a. cold
    b. occluded
    c. stationary
    d. warm

    ANS: A   DIF: I   OBJ: 2-2.4

58. _____ usually bring drizzly precipitation. Afterward, weather conditions are clear and warm.
    a. Cold fronts
    b. Stationary fronts
    c. Occluded fronts
    d. Warm fronts

    ANS: D   DIF: I   OBJ: 2-2.4

59. A(n) _____ has cool temperatures and large amounts of precipitation.
    a. stationary front
    b. cold front
    c. occluded front
    d. warm front

    ANS: C   DIF: I   OBJ: 2-2.4

60. The weather associated with a(n) _____ is similar to that produced by a warm front.
    a. occluded front
    b. cold front
    c. stationary front
    d. massive front

    ANS: C   DIF: I   OBJ: 2-2.4

Below are four statements describing the formation of tornadoes. However, they are all out of order. Read the statements below and answer the question that follows.
A) The rotating column of air works its way down to the bottom of the cumulonimbus cloud and forms a funnel cloud.
B) The rotating column of air is turned to a vertical position by strong updrafts of air within the cumulonimbus cloud. The updrafts of air also begin to rotate with the column of air.
C) Wind traveling in two different directions causes a layer of air in the middle to begin to rotate like a roll of toilet paper.
D) The funnel cloud touches the ground.

61. In what order should the statements above appear to correctly describe how a tornado forms?
    a. A, B, C, D
    b. B, C, A, D
    c. C, B, A, D
    d. D, A, B, C

    ANS: C   DIF: II   OBJ: 2-3.2

62. Which device is used to measure air temperature?
    a. barometer
    b. anemometer
    c. thermometer
    d. windsock or wind vane

    ANS: C   DIF: I   OBJ: 2-4.1

63. Which device is used to measure air pressure?
    a. barometer
    b. thermometer
    c. anemometer
    d. windsock or wind vane

    ANS: A   DIF: I   OBJ: 2-4.1

64. Which device is used to measure wind speed?
    a. thermometer
    b. anemometer
    c. barometer
    d. windsock or wind vane

    ANS: B   DIF: I   OBJ: 2-4.1

65. Which device is used to measure wind direction?
    a. barometer
    b. anemometer
    c. thermometer
    d. windsock or wind vane

    ANS: D   DIF: I   OBJ: 2-4.1

66. When *-nimbus* or *nimbo-* is part of a cloud's name, it means that
    a. the cloud is at a high altitude.
    b. it is a middle cloud.
    c. precipitation might fall from the cloud.
    d. it is a low cloud.

    ANS: C   DIF: I   OBJ: 2-1.4

67. When *cirro-* is part of a cloud's name, it means that
    a. the cloud is at a high altitude.
    b. it is a low cloud.
    c. precipitation might fall from the cloud.
    d. it is a middle cloud.

    ANS: A   DIF: I   OBJ: 2-1.4

68. When *alto-* is part of a cloud's name, it means that
    a. the cloud is at a high altitude.   c. it is a low cloud.
    b. precipitation might fall from the cloud   d. it is a middle cloud.

    ANS: D   DIF: I   OBJ: 2-1.4

69. When *strato-* is part of a cloud's name, it means that
    a. it is a low cloud.   c. the cloud is at a high altitude.
    b. precipitation might fall from the cloud   d. it is a middle cloud.

    ANS: A   DIF: I   OBJ: 2-1.4

70. What type of clouds would you most likely see if you were flying in an airplane at 8,000 m?
    a. cumulus clouds   c. cirrus clouds
    b. stratus clouds   d. altonimbus clouds

    ANS: C   DIF: II   OBJ: 2-1.4

71. Suppose that 1 m^3 of air at a certain temperature can hold 10 g of water vapor. However, you know that the air actually contains 9 g of water vapor. What is the relative humidity?
    a. 30 percent   c. 75 percent
    b. 60 percent   d. 90 percent

    ANS: D   DIF: II   OBJ: 2-1.2

COMPLETION

1. One can often see the shapes of animals and people in fluffy, white _____ clouds. (cumulus or stratus)

   ANS: cumulus   DIF: I   OBJ: 2-1.4

2. Wind speed can be measured using a(n) _____. (windsock or anemometer)

   ANS: anemometer   DIF: I   OBJ: 2-4.1

3. People use coasters when setting cold drinks on furniture to protect the surface from _____. (precipitation or condensation)

   ANS: condensation   DIF: I   OBJ: 2-1.3

4. Fog is a type of _____ cloud. (cirrus or stratus)

   ANS: stratus DIF: I OBJ: 2-1.4

5. A psychrometer measures _____. (humidity or relative humidity)

   ANS: relative humidity DIF: I OBJ: 2-1.2

6. The terms occluded and stationary describe types of _____. (fronts or air masses)

   ANS: fronts DIF: I OBJ: 2-2.3

7. _____ is the condition of the atmosphere at a particular time and place.

   ANS: Weather DIF: I OBJ: 2-1.1

8. The _____ is the continuous movement of water from water sources, such as lakes and oceans, into the air, onto and over land, into the ground, and back to the water sources.

   ANS: water cycle DIF: I OBJ: 2-1.1

9. When air holds all the water it can at a given temperature, the air is said to be _____.

   ANS: saturated DIF: I OBJ: 2-1.2

10. The _____ is the temperature to which air must cool to be completely saturated.

    ANS: dew point DIF: I OBJ: 2-1.3

11. A(n) _____ is a large body of air that has similar temperature and moisture throughout.

    ANS: air mass DIF: I OBJ: 2-2.1

12. When two different air masses meet, a boundary forms between them called a _____.

    ANS: front DIF: I OBJ: 2-2.3

13. _____ are small, intense weather systems that produce strong winds, heavy rain, lightning, and thunder.

    ANS: Thunderstorms DIF: I OBJ: 2-3.2

14. _____ is the sound that results from the rapid expansion of air along the lightning strike.

    ANS: Thunder   DIF: I   OBJ: 2-3.1

15. _____ is a large electrical discharge that occurs between two oppositely charged surfaces.

    ANS: Lightning   DIF: I   OBJ: 2-3.1

16. A(n) _____ is a small, rotating column of air that has high wind speeds and low central pressure and that touches the ground.

    ANS: tornado   DIF: I   OBJ: 2-3.3

17. A(n) _____ is a large, rotating tropical weather system with wind speeds of at least 119 km/h.

    ANS: hurricane   DIF: I   OBJ: 2-3.3

18. The _____ is the center of the hurricane that is a core of warm, relatively calm air with low pressure and light winds.

    ANS: eye   DIF: I   OBJ: 2-3.3

19. The strongest part of a hurricane is called the _____. This is a group of clouds that produce heavy rains and forceful winds that can reach speeds of 300 km/h.

    ANS: eye wall   DIF: I   OBJ: 2-3.3

20. Beyond the eye wall, spiraling bands of clouds called _____ circle the center of the hurricane, producing heavy rains and high winds.

    ANS: rain bands   DIF: I   OBJ: 2-3.3

21. A _____ is a prediction of weather conditions over the next three to five days.

    ANS: weather forecast   DIF: I   OBJ: 2-4.1

22. _____ is used to find the location, movement, and intensity of precipitation as well as detect what form of precipitation a weather system is carrying.

    ANS: Radar   DIF: I   OBJ: 2-4.1

23. _____ orbiting the Earth provide the images of the swirling clouds you can see on television weather reports. They can measure wind speeds, humidity, and the temperatures at various altitudes.

   ANS: Weather satellites   DIF: I   OBJ: 2-4.1

24. Similar to contour lines on a topographical map, _____ are lines that connect points of equal air pressure rather than equal elevation.

   ANS: isobars   DIF: I   OBJ: 2-4.2

25. A _____ is a collection of millions of tiny water droplets or ice crystals.

   ANS: cloud   DIF: I   OBJ: 2-1.4

26. _____ occurs when saturated air cools further.

   ANS: Condensation   DIF: I   OBJ: 2-1.3

27. A _____ is an instrument used to measure the amount of rainfall and typically consists of a funnel and a cylinder.

   ANS: rain gauge   DIF: I   OBJ: 2-1.5

28. When rain does NOT freeze until it hits a surface near the ground, a _____, or layer of ice, forms.

   ANS: glaze   DIF: I   OBJ: 2-1.5

SHORT ANSWER

For each pair of terms, explain the difference in their meanings.

1. barometer/anemometer

   ANS:
   A barometer measures air pressure. An anemometer measures wind speed.

   DIF: I   OBJ: 2-4.1

2. tornado/hurricane

   ANS:
   A tornado is a small, rotating column of air with high wind speed that touches the ground. A hurricane is a large, rotating tropical weather system with wind speeds equal to or greater than 119 km/h.

   DIF: I   OBJ: 2-3.2

3. lightning/thunder

   ANS:
   Lightning is a large electrical discharge that occurs between two oppositely charged surfaces. Thunder is the sound that results from the rapid expansion of air along a lightning strike.

   DIF: I   OBJ: 2-3.1

4. air mass/front

   ANS:
   An air mass is a large body of air that has the same moisture and temperature throughout. A front is the boundary that forms where two different air masses meet.

   DIF: I   OBJ: 2-2.3

5. condensation/precipitation

   ANS:
   Condensation is the process by which a gas changes state to become a liquid. Precipitation is liquid or solid water that falls from the atmosphere to Earth.

   DIF: I   OBJ: 2-1.1

6. relative humidity/dew point

   ANS:
   Relative humidity is the amount of water vapor the air contains relative to the maximum amount it can hold at a given temperature. Dew point is the temperature to which air must cool to be saturated.

   DIF: I   OBJ: 2-1.2

7. What is the difference between humidity and relative humidity?

   ANS:
   Humidity is the amount of water vapor in the air. Relative humidity is the amount of water vapor the air contains compared with the maximum amount it can hold at a given temperature.

   DIF: I   OBJ: 2-1.2

8. What are two ways that air can become saturated with water vapor?

   ANS:
   Air can become saturated if water evaporates into the air or if the air temperature drops.

   DIF: I   OBJ: 2-1.3

9. What does a relative humidity of 75 percent mean?

   ANS:
   The air is holding 75 percent of the amount of water it can hold at a given temperature.

   DIF: I   OBJ: 2-1.2

10. How does the water cycle contribute to condensation?

    ANS:
    Before condensation can occur, the air must be saturated. Evaporation, a part of the water cycle, adds water to the air.

    DIF: I   OBJ: 2-1.1

11. What happens to relative humidity as the air temperature drops below the dew point?

    ANS:
    As the air temperature drops below the dew point, relative humidity increases to the point that the air becomes saturated with moisture and condensation occurs.

    DIF: II   OBJ: 2-1.2

12. How do clouds form?

    ANS:
    Clouds form as warm air rises and cools. As the air cools, it becomes saturated. If there are condensation surfaces available, the water vapor changes physical states to liquid droplets or solid ice crystals, forming a cloud.

    DIF: I   OBJ: 2-1.4

13. Why are some clouds formed from water droplets, while others are made up of ice crystals?

    ANS:
    At higher temperatures, water vapor condenses on surfaces as tiny water droplets. When temperatures are below freezing, water vapor changes directly to ice crystals.

    DIF: I   OBJ: 2-1.4

14. Describe how rain forms.

    ANS:
    Rain forms when a cloud's water droplets become too heavy to remain suspended in the cloud. The droplets grow by colliding and joining with other droplets.

    DIF: I   OBJ: 2-1.5

15. How can rain and hail fall from the same cumulonimbus cloud?

    ANS:
    Hail forms when raindrops are carried by updrafts of air to higher altitudes in clouds, where the raindrops freeze. Some raindrops may not be caught in the updrafts and will fall to the ground as rain.

    DIF: II   OBJ: 2-1.5

16. What are the characteristics that define air masses?

    ANS:
    Temperature and moisture are the characteristics that define air masses.

    DIF: I   OBJ: 2-2.1

17. What are the major air masses that influence the weather in the United States?

    ANS:
    The major air masses that influence the weather in the United States are maritime tropical, maritime polar, continental tropical, and continental polar.

    DIF: I   OBJ: 2-2.2

18. a. What are fronts?
    b. What causes fronts?

    ANS:
    a. Fronts are boundaries that form between two different air masses.
    b. Boundaries form because air masses with different moisture and temperature characteristics do not mix easily.

    DIF: I   OBJ: 2-2.3

19. What kind of front forms when a cold air mass displaces a warm air mass?

    ANS:
    A cold front forms.

    DIF: I   OBJ: 2-2.3

20. Explain why the Pacific Coast has cool, wet winters and warm, dry summers.

    ANS:
    In the winter, the Pacific Coast's climate is governed by a maritime polar air mass that brings wet weather and cool temperatures. In the summer, the Pacific Coast's climate is governed by a maritime tropical air mass that brings warm temperatures and little moisture.

    DIF: II   OBJ: 2-2.4

21. What is lightning?

    ANS:
    Lightning is a large electrical discharge that occurs between two oppositely charged surfaces.

    DIF: I   OBJ: 2-3.1

22. a. Describe how tornadoes develop.
    b. What is the difference between a funnel cloud and a tornado?

    ANS:

    a. A tornado develops when wind traveling in two different directions causes the air in the middle to rotate. The rotating column of air is turned upright by updrafts that begin spinning with it. The rotating air works its way down to the bottom of the cloud and forms a funnel cloud.
    b. When the funnel cloud touches the ground, it is called a tornado.

    DIF: I   OBJ: 2-3.2

23. Why do hurricanes form only over certain areas?

    ANS:
    Hurricanes form only over warm, tropical oceans because a hurricane requires the energy and moisture from water to fuel it.

    DIF: I   OBJ: 2-3.2

24. What happens to a hurricane as it moves over land? Why?

    ANS:
    A hurricane dissipates as it moves over land because it loses its energy source.

    DIF: II   OBJ: 2-3.3

25. What are three methods meteorologists use to collect weather data?

    ANS:
    Answers will vary. Sample answer: weather balloons, Doppler radar, and weather satellites.

    DIF: I   OBJ: 2-4.1

26. What are weather maps based on?

    ANS:
    Weather maps are based on weather data gathered from weather stations across the United States.

    DIF: I   OBJ: 2-4.2

27. What does a station model represent?

    ANS:
    A station model represents the location of the weather station and the weather data collected there.

    DIF:　I　　　　　OBJ:　2-4.2

28. Why would a meteorologist compare a new weather map with one 24 hours old?

    ANS:
    Answers may vary. Sample answer: Meteorologists would compare a new weather map with one 24 hours old to see how fast a front is moving.

    DIF:　II　　　　OBJ:　2-4.2

29. Compare and contrast the processes of condensation and evaporation in the water cycle.

    ANS:
    Both are processes in Earth's water cycle that involve a change of the state of water. They differ in that *condensation* occurs when water vapor changes to a liquid, while *evaporation* occurs when liquid water changes to a gas.

    DIF:　I　　　　　OBJ:　2-1.1

30. What name would you give a lacy, layered cloud above 6,000 m?

    ANS:
    Cirrostratus clouds are a lacy, layered clouds above 6,000 m.

    DIF:　I　　　　　OBJ:　2-1.4

31. Compare and contrast snow, sleet, and hail.

    ANS:
    All are forms of solid precipitation that fall from clouds. *Snow* forms when water vapor changes to a solid. *Sleet* forms when rain falls through a layer of freezing air. *Hail* forms when raindrops are carried by winds to higher altitudes in a cloud, where they freeze and accumulate layers.

    DIF:　I　　　　　OBJ:　2-1.5

32. If a continental polar air mass moves over Ohio in the summer, what will the weather be like?

    ANS:
    The weather will probably be cool and dry.

    DIF:　I　　　　　OBJ:　2-2.2

33. Why does the continental tropical air mass that forms over northern Mexico bring clear, dry, hot weather?

    ANS:
    It forms over the desert, which is hot and contains relatively little moisture.

    DIF: I   OBJ: 2-2.2

34. Explain how a cold front develops.

    ANS:
    A cold front develops when a cold air mass moves under a warm air mass, forcing the warmer air upward.

    DIF: I   OBJ: 2-2.3

35. What kind of weather is associated with a stationary front?

    ANS:
    It will probably be cloudy and rainy as long as the front lies over an area. After the front passes, the weather will usually clear up.

    DIF: I   OBJ: 2-2.4

36. What is the relationship between lightning and thunder?

    ANS:
    Lightning is an electrical discharge that forms between clouds or between a cloud and the ground. The air around the lightning bolt expands rapidly, producing sound waves that we call *thunder*.

    DIF: I   OBJ: 2-3.1

37. Explain why tornadoes often destroy buildings in their path.

    ANS:
    Buildings are often destroyed by the enormous force exerted by tornado winds and by the strong updrafts that accompany them.

    DIF: I   OBJ: 2-3.3

38. Why don't hurricanes form over land?

    ANS:
    A hurricane gets it energy from enormous volumes of warm, moist air, which are not present over land masses.

    DIF: I   OBJ: 2-3.2

Holt Science and Technology
Copyright © by Holt, Rinehart and Winston. All rights reserved.

39. Would water be a useful fluid to use in a thermometer? Explain.

    ANS:
    No; water would not be a good thermometer fluid because it expands when it freezes.

    DIF: I   OBJ: 2-4.1

40. What advantage do weather satellites have over ground-based weather stations?

    ANS:
    Satellites can gather weather data from much higher altitudes than land-based instruments can.

    DIF: I   OBJ: 2-4.1

41. Why are so many station models used to gather weather data in the United States?

    ANS:
    Because the country is so large, and Earth's atmosphere is constantly changing, we need data from many stations to make accurate forecasts.

    DIF: I   OBJ: 2-4.2

42. Explain the relationship between condensation and dew point.

    ANS:
    The air must cool to below its dew point before condensation can occur.

    DIF: I   OBJ: 2-1.3

43. Describe the conditions along a stationary front.

    ANS:
    Stationary fronts generally bring drizzly precipitation. After the front passes, the weather is generally clear and warm.

    DIF: I   OBJ: 2-2.4

44. What are the characteristics of an air mass that forms over the Gulf of Mexico?

    ANS:
    An air mass that forms over the Gulf of Mexico is warm and wet.

    DIF: I   OBJ: 2-2.2

45. Explain how a hurricane develops.

    ANS:
    A hurricane begins as a group of thunderstorms moving over tropical ocean waters. Winds traveling in two different directions collide, causing the storm to rotate over an area of low pressure. The hurricane is fueled by the condensation of water vapor.

    DIF: I   OBJ: 2-3.2

46. Use the following terms to create a concept map: *evaporation, relative humidity, water vapor, dew, psychrometer, clouds, fog.*

    ANS:

    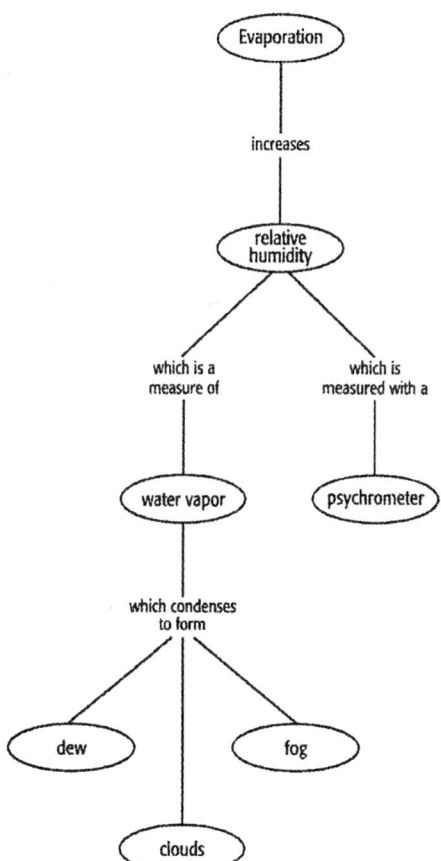

    DIF: II   OBJ: 2-1.5

47. If both the air temperature and the amount of water vapor in the air change, is it possible for the relative humidity to stay the same? Explain.

    ANS:
    Yes; for example, if both the air temperature and water vapor increased, the relative humidity might remain the same.

    DIF: II   OBJ: 2-1.2

48. What can you assume about the amount of water vapor in the air if there is no difference between the wet- and dry-bulb readings of a psychrometer?

    ANS:
    The air is saturated with water.

    DIF: II   OBJ: 2-1.2

49. List the major similarities and differences between hurricanes and tornadoes.

    ANS:
    Answers may vary. Sample answer: Both begin as a result of thunderstorms and are centered around low pressure. Hurricanes occur over water, and tornadoes generally occur over land.

    DIF: II   OBJ: 2-3.3

    You always see lightning before you hear thunder. That's because light travels at about 300,000,000 m/s, while sound travels only 330 m/s. One way you can determine how close you are to the thunderstorm is by counting how many seconds there are between the lightning and thunder. Usually, it takes thunder about 3 seconds to cover 1 km. Answer the following questions based on this estimate:

50. If you hear thunder 12 seconds after you see the flash of lightning, how far away is the thunderstorm?

    ANS:
    (12 s ÷ 3 s) × 1 km = 4 km

    DIF: II   OBJ: 2-3.1

51. If you hear thunder 36 seconds after you see the flash of lightning, how far away is the thunderstorm?

    ANS:
    (36 s ÷ 3 s) × 1 km = 12 km

    DIF: II   OBJ: 2-3.1

Use the weather map below to answer the questions that follow.

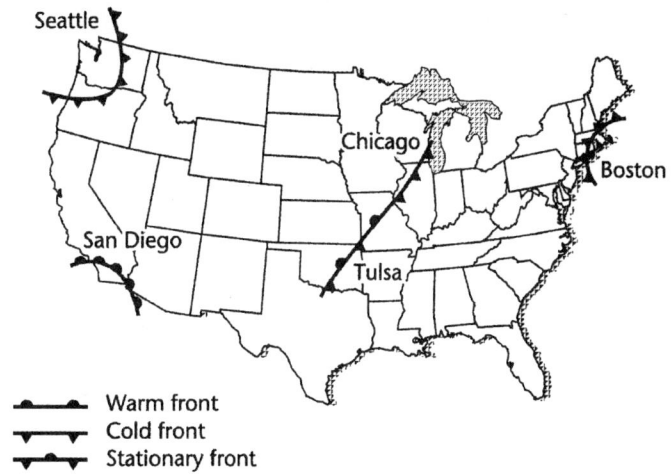

52. Where are thunderstorms most likely to occur? Explain your answer.

    ANS:
    Thunderstorms are most likely to occur in Chicago because a cold front is approaching.

    DIF: II   OBJ: 2-4.2

53. What are the weather conditions like in Tulsa, Oklahoma? Explain your answer.

    ANS:
    Tulsa is experiencing a stationary front. It is probably receiving drizzly precipitation.

    DIF: II   OBJ: 2-4.2

54. What causes tornadoes? Explain.

    ANS:
    Wind traveling in opposite directions cause a layer of air in the middle to begin to rotate. The rotating layer of air changes from a horizontal to a vertical position due to strong updrafts which join the rotational pattern. The rotating air column moves to the bottom of the cumulonimbus cloud and forms a funnel cloud. When the cloud touches the ground, it is called a tornado.

    DIF: I   OBJ: 2-3.2

55. What causes lightning? Explain.

    ANS:
    The upper part of a cloud usually has a positive charge, and the lower part of a cloud usually has a negative charge. Lightning is formed when the opposite charges on the base of one cloud and another cloud or the ground cause a large electrical discharge.

    DIF: I   OBJ: 2-3.1

Holt Science and Technology
Copyright © by Holt, Rinehart and Winston. All rights reserved.

56. Use the following terms to complete the concept map below: *polar, source regions, warm air, maritime, dry air, temperature, moisture.*

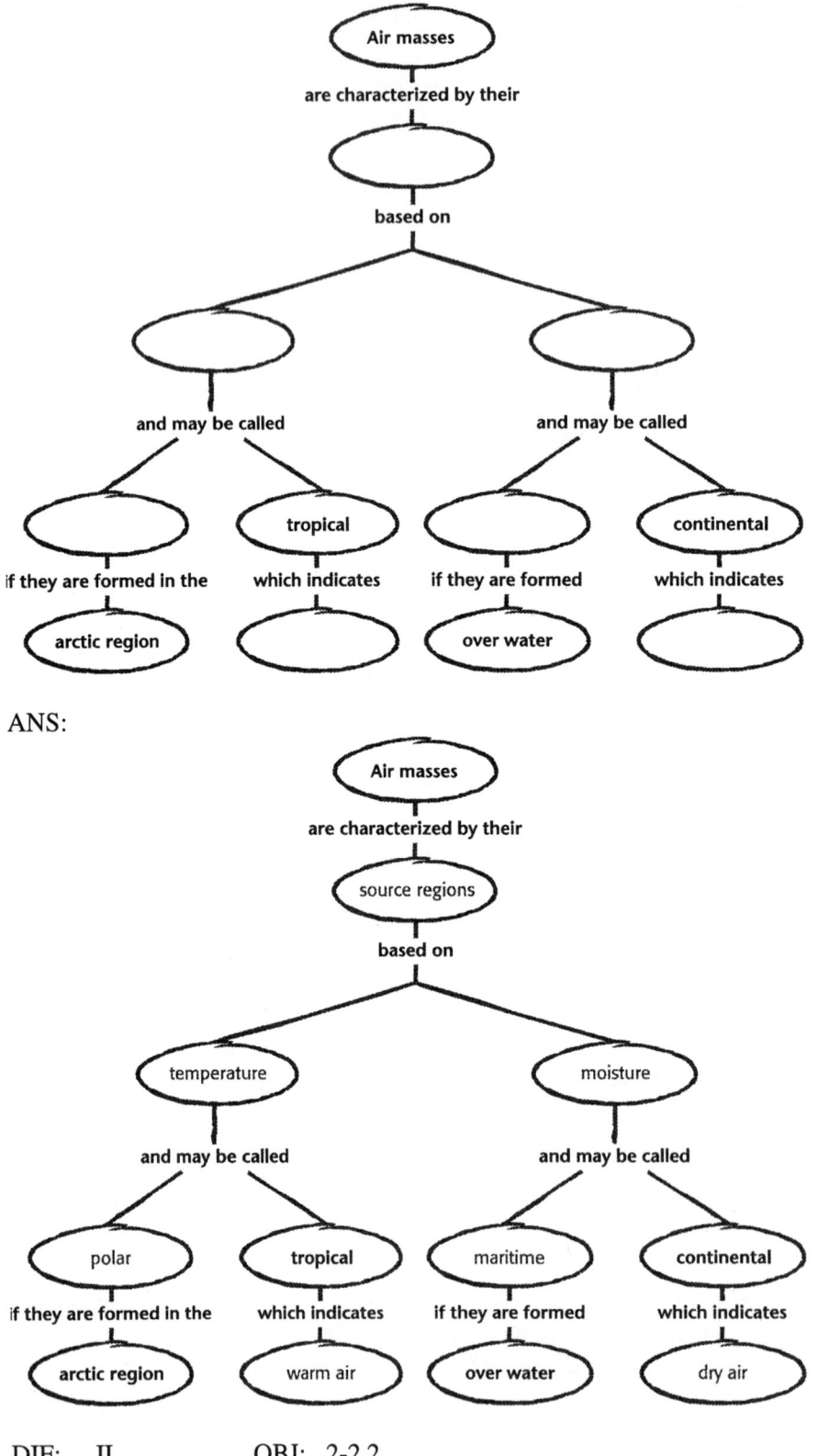

ANS:

DIF: II   OBJ: 2-2.2

57. Dew is formed when small water droplets condense on grass. In what type of environment is dew NOT likely to form?

   ANS:
   Sample answer: In order for water droplets to condense, the air must have a relative humidity of 100 percent. In areas that are very dry, such as a desert, there is not enough humidity to form dew.

   DIF: II   OBJ: 2-1.3

   Examine the graph below and answer the questions that follow.

   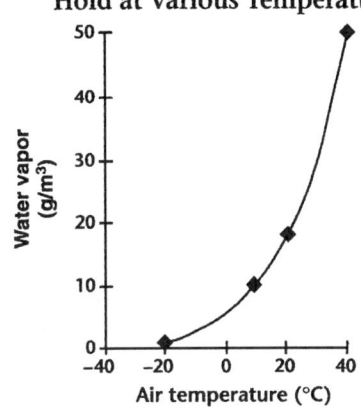

58. About how much moisture can air hold at 20°C?

   ANS:
   At 20°C, air can hold about 18 g/m3 of water.

   DIF: II   OBJ: 2-1.3

59. What does this graph tell you about the relationship between temperature and the amount of water vapor that air can hold?

   ANS:
   The amount of water vapor that air can hold increases as temperature increases.

   DIF: II   OBJ: 2-1.3

Short Course I Chapter 3—Climate

MULTIPLE CHOICE

1. The tilt of Earth as it orbits around the sun causes
   a. global warming
   b. different seasons.
   c. a rain shadow.
   d. the greenhouse effect.

   ANS: B   DIF: I   OBJ: 3-1.2

2. What factor affects the prevailing winds as they blow across a continent, producing different climates?
   a. latitude
   b. mountains
   c. forests
   d. glaciers.

   ANS: B   DIF: I   OBJ: 3-1.2

3. What factor determines the amount of solar energy an area receives?
   a. latitude
   b. wind patterns
   c. mountains
   d. ocean currents

   ANS: A   DIF: I   OBJ: 3-1.2

4. What climate zone has the coldest average temperature?
   a. tropical
   b. polar
   c. temperate
   d. tundra

   ANS: B   DIF: I   OBJ: 3-2.1

5. What biome is not located in the tropical zone?
   a. rain forest
   b. savanna
   c. chaparral
   d. desert

   ANS: C   DIF: I   OBJ: 3-2.2

6. What biome contains the greatest number of plant and animal species?
   a. rain forest
   b. temperate forest
   c. grassland
   d. tundra

   ANS: A   DIF: I   OBJ: 3-2.2

7. Which of the following is NOT a theory for the cause of ice ages?
   a. the Milankovich theory
   b. volcanic eruptions
   c. plate tectonics
   d. the greenhouse effect

   ANS: D   DIF: I   OBJ: 3-3.2

8. Which of the following is thought to contribute to global warming?
   a. wind patterns
   b. deforestation
   c. ocean surface currents
   d. microclimates

   ANS: B   DIF: I   OBJ: 3-3.3

Holt Science and Technology
Copyright © by Holt, Rinehart and Winston. All rights reserved.

9. Wind will generally carry the most moisture when it comes from
   a. warm grasslands.
   b. warm tropical seas.
   c. polar icecaps.
   d. mountainous regions.

   ANS: B   DIF: I   OBJ: 3-1.2

10. Which of these gets about as much precipitation as a tropical desert?
    a. taiga
    b. tundra
    c. tropical rain forest
    d. chaparral

    ANS: B   DIF: I   OBJ: 3-2.2

11. Which human activity is thought to be most responsible for global warming?
    a. using aerosol hairspray
    b. mowing grasslands for cattle feed
    c. burning fossil fuels for transportation
    d. planting deciduous forests for building materials

    ANS: C   DIF: I   OBJ: 3-3.3

12. Seasons are caused by
    a. the gravitational pull of the moon.
    b. centripetal force.
    c. sunspots.
    d. the tilt of the Earth's axis.

    ANS: D   DIF: I   OBJ: 3-1.2

13. The equator has a latitude of
    a. 0°.
    b. 45°.
    c. 90°.
    d. 180°.

    ANS: A   DIF: I   OBJ: 3-1.2

14. The North Pole has a latitude of
    a. 0°.
    b. 45°.
    c. 90°.
    d. 180°.

    ANS: C   DIF: I   OBJ: 3-1.2

15. At a latitude of 90°, the sun's rays strike the surface of the Earth at a
    a. 90° angle spreading the amount of solar energy over a large area.
    b. 90° angle concentrating the amount of solar energy onto a small area.
    c. lesser angle concentrating the amount of solar energy onto a small area.
    d. lesser angle spreading the amount of solar energy over a large area.

    ANS: D   DIF: I   OBJ: 3-1.2

Holt Science and Technology
Copyright © by Holt, Rinehart and Winston. All rights reserved.

16. At the equator, the sun's rays strike the surface of the Earth at a
    a. 90° angle spreading the amount of solar energy over a large area.
    b. 90° angle concentrating the amount of solar energy onto a small area.
    c. lesser angle concentrating the amount of solar energy onto a small area.
    d. lesser angle spreading the amount of solar energy over a large area.

    ANS: B　　　　　DIF: I　　　　　OBJ: 3-1.2

17. The polar regions receive almost 24 hours of darkness in the
    a. summer.　　　　　　　c. fall.
    b. winter.　　　　　　　d. spring.

    ANS: B　　　　　DIF: I　　　　　OBJ: 3-1.2

18. The polar regions receive almost 24 hours of daylight in the
    a. summer.　　　　　　　c. fall.
    b. spring.　　　　　　　d. winter.

    ANS: A　　　　　DIF: I　　　　　OBJ: 3-1.2

19. During our winter months, the Southern Hemisphere has
    a. lower temperatures and shorter days.　　c. higher temperatures and longer days.
    b. lower temperatures and longer days.　　d. higher temperatures and shorter days.

    ANS: C　　　　　DIF: I　　　　　OBJ: 3-1.2

20. Because warm air is
    a. less dense, it tends to rise.　　　c. more dense, it tends to rise.
    b. less dense, it tends to sink.　　　d. more dense, it tends to sink.

    ANS: A　　　　　DIF: I　　　　　OBJ: 3-1.2

21. If prevailing winds form from cold air, they will probably
    a. be dry.　　　　　　　c. produce sleet.
    b. produce rain.　　　　　d. produce snow.

    ANS: A　　　　　DIF: I　　　　　OBJ: 3-1.2

22. Even though Mt. Kilimanjaro is only about 3° south of the equator, it is snow-covered year-round. This is because of
    a. its latitude.　　　　　c. its elevation.
    b. surface currents.　　　d. wind patterns.

    ANS: C　　　　　DIF: I　　　　　OBJ: 3-1.2

23. As elevation increases, atmosphere becomes
    a. more dense and temperature decreases.
    b. less dense and temperature decreases.
    c. less dense and temperature increases.
    d. more dense and temperature increases.

    ANS: B　　　　　DIF: I　　　　　OBJ: 3-1.2

24. If a prevailing wind blows from an ocean in the west across a mountain in the east, the land on the windward side of the mountain
    a. and the land on the other side will be dry.
    b. and the land on the other side will be green and lush.
    c. will be dry and the land on the other side will be green and lush.
    d. will be green and lush and the land on the other side will be dry.

    ANS: D   DIF: I   OBJ: 3-1.2

25. If one side of a mountain resembles a dry desert, we call this a
    a. tundra.   c. rain shadow.
    b. taiga.   d. polar zone.

    ANS: C   DIF: I   OBJ: 3-1.2

26. Iceland and Greenland are both countries just below the Arctic Circle. Why does Iceland have a warmer climate than Greenland?
    a. elevation   c. latitude
    b. surface currents   d. precipitation

    ANS: B   DIF: I   OBJ: 3-1.2

Examine the diagram of an ocean current and answer the question that follows.

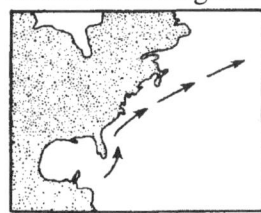

27. What is the name of the current in the illustration above?
    a. Labrador Current   c. Gulf Stream
    b. North Atlantic Drift   d. North Equatorial Current

    ANS: C   DIF: II   OBJ: 3-1.2

28. Which of the following is NOT true?
    a. The Gulf Stream is a warm current.
    b. Continental land masses act as barriers to surface currents.
    c. Continental land masses act as barriers to prevailing winds.
    d. The surface temperature of water affects the temperature of the air above it.

    ANS: C   DIF: I   OBJ: 3-1.2

Examine the map below and answer the question that follows.

29. Which statement does NOT correctly describe surface currents?
    a. Surface currents along the Antarctic move eastward.
    b. Surface currents are so short that they do not affect global climates.
    c. Surface currents in the Southern Hemisphere move counterclockwise.
    d. Surface currents in the Northern Hemisphere move clockwise.

    ANS: B   DIF: II   OBJ: 3-1.2

30. Which of the following is a biome of a major climate zone?
    a. tundra   c. polar
    b. tropical   d. temperate

    ANS: A   DIF: I   OBJ: 3-2.2

31. Which of the following is NOT a major climate zone?
    a. polar   c. antarctic
    b. tropical   d. temperate

    ANS: C   DIF: I   OBJ: 3-2.1

Below is an illustration in which the Earth has been divided into the major climate zones. Examine the illustration and answer the questions that follow.

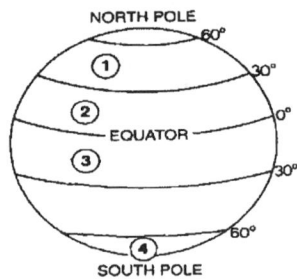

32. Which of the following is a climate zone at **3**?
    a. polar   c. antarctic
    b. tropical   d. temperate

    ANS: B   DIF: II   OBJ: 3-2.1

33. Which of the following is a climate zone at **1**?
    a. polar   c. antarctic
    b. tropical   d. temperate

    ANS: D   DIF: II   OBJ: 3-2.1

Holt Science and Technology
Copyright © by Holt, Rinehart and Winston. All rights reserved.

34. Which of the following is a climate zone at **4**?
    a. polar
    b. tropical
    c. antarctic
    d. temperate

    ANS: A   DIF: II   OBJ: 3-2.1

35. Climate zones are determined by
    a. geography.
    b. precipitation amounts.
    c. temperature ranges.
    d. Both (a) and (b)

    ANS: C   DIF: I   OBJ: 3-2.1

36. Biomes are determined by
    a. geography.
    b. precipitation amounts.
    c. temperature ranges.
    d. All of the above

    ANS: D   DIF: I   OBJ: 3-2.2

37. Which biome is NOT in the temperate zone?
    a. taiga
    b. desert
    c. forest
    d. chaparral

    ANS: A   DIF: I   OBJ: 3-2.2

38. Which biome is in the polar zone?
    a. forest
    b. tundra
    c. chaparral
    d. grasslands

    ANS: B   DIF: I   OBJ: 3-2.2

39. Which of the following is a succulent plant?
    a. cactus
    b. fern
    c. herb
    d. thorny shrub

    ANS: A   DIF: I   OBJ: 3-2.2

40. The seeds of some savanna plants require ____ to grow.
    a. no water
    b. fertilizer
    c. fire
    d. darkness

    ANS: C   DIF: I   OBJ: 3-2.2

41. In which biome would you expect to find a kangaroo rat?
    a. tundra
    b. taiga
    c. chaparral
    d. tropical desert

    ANS: D   DIF: I   OBJ: 3-2.2

Holt Science and Technology
Copyright © by Holt, Rinehart and Winston. All rights reserved.

42. Prairies, steppes, and pampas are all local names of
    a. tropical savannas.
    b. temperate grasslands.
    c. tundras.
    d. tropical deserts.

    ANS: B　　　DIF: I　　　OBJ: 3-2.2

43. Most of the ____ have been plowed to make room for croplands.
    a. chaparrals
    b. temperate grasslands
    c. tropical savannas
    d. temperate forests

    ANS: B　　　DIF: I　　　OBJ: 3-2.2

44. Which biome tends to be very hot in the daytime and very cold at night?
    a. tundra
    b. tropical desert
    c. tropical rain forest
    d. temperate desert

    ANS: D　　　DIF: I　　　OBJ: 3-2.2

45. The ____ is the biome that provides most of the wood for paper.
    a. tundra
    b. tropical rain forest
    c. taiga
    d. chaparral

    ANS: C　　　DIF: I　　　OBJ: 3-2.2

46. A large city that is 2°C warmer than the surrounding rural areas is an example of a
    a. climate zone.
    b. microclimate.
    c. biome.
    d. tropical zone.

    ANS: B　　　DIF: I　　　OBJ: 3-2.2

47. Which biome has the most fertile soil?
    a. tropical rain forest
    b. tropical savanna
    c. tundra
    d. temperate grassland

    ANS: D　　　DIF: I　　　OBJ: 3-2.2

48. In which biome would you most likely find monkeys, lemurs, and jaguars?
    a. tropical rain forest
    b. tropical savanna
    c. chaparral
    d. temperate forest

    ANS: A　　　DIF: I　　　OBJ: 3-2.2

49. During a glacial period, large amounts of ocean water
    a. freeze and the sea levels drop.
    b. freeze and the sea levels rise.
    c. melt and the sea levels drop.
    d. melt and the sea levels rise.

    ANS: A　　　DIF: I　　　OBJ: 3-3.1

50. During an interglacial period, large amounts of ocean water
    a. freeze and the sea levels drop.
    b. freeze and the sea levels rise.
    c. melt and the sea levels drop.
    d. melt and the sea levels rise.

    ANS: D   DIF: I   OBJ: 3-3.1

51. Milutin Milankovitch, a Yugoslavian scientist, suggested that over a period of 100,000 years the shape of Earth's orbit changes. This would cause
    a. hotter summers and colder winters when the orbit is circular.
    b. hotter summers and colder winters when the orbit is elliptical.
    c. cooler summers and warmer winters when the orbit is elliptical.
    d. hotter summers and colder winters when the orbit is circular.

    ANS: B   DIF: I   OBJ: 3-3.2

52. Milutin Milankovitch suggested that the tilt of the Earth varies between 21.8° and 24.4°. When the tilt is at
    a. 24.4°, the poles receive more solar energy.
    b. 24.4°, the poles receive less solar energy.
    c. 21.8°, the poles receive more solar energy.
    d. Both (a) and (c)

    ANS: A   DIF: I   OBJ: 3-3.2

53. How do volcanic eruptions, specifically the worldwide spread of volcanic ash during an eruption, affect global climate?
    a. Volcanic ash absorbs the sun's rays, causing cooler temperatures.
    b. Volcanic ash intensifies the sun's rays like a magnifying glass, warming the Earth.
    c. Volcanic ash reflects the sun's rays, causing cooler temperatures.
    d. Volcanic ash affects only the local climate by destroying everything.

    ANS: C   DIF: I   OBJ: 3-3.2

54. Evidence suggests that one reason Earth's climate has changed might be because all continents were once
    a. near the South Pole.
    b. near the North Pole.
    c. near the equator.
    d. split into two groups, one at each pole.

    ANS: A   DIF: I   OBJ: 3-3.2

55. One theory suggests that all of the continents were once a giant landmass called
    a. Pangaea.
    b. Panthalassa.
    c. Laurasia.
    d. Gondwana.

    ANS: A   DIF: I   OBJ: 3-3.2

56. Which gas is thought to contribute to global warming?
    a. nitrogen
    b. carbon monoxide
    c. oxygen
    d. carbon dioxide

    ANS: D   DIF: I   OBJ: 3-3.3

57. One likely consequence of global warming would be
    a. earthquakes.
    b. volcanic eruptions.
    c. flooding.
    d. tornadoes.

    ANS: C   DIF: I   OBJ: 3-3.3

58. Where would the weather most likely be warm on December 25th?
    a. Alaska
    b. Australia
    c. England
    d. Canada

    ANS: B   DIF: II   OBJ: 3-1.2

59. In Africa, the Sahara desert is near the west coast of the continent. Why, when it is so near to the ocean, does it NOT receive much precipitation?
    a. Surface currents cause it to be too hot.
    b. Prevailing winds flow from west to east.
    c. Prevailing winds flow from east to west.
    d. Its elevation is too high above sea level.

    ANS: C   DIF: II   OBJ: 3-1.2

60. As air is forced up over a mountain, it
    a. cools and releases moisture.
    b. cools and absorbs moisture.
    c. warms and releases moisture.
    d. warms and absorbs moisture.

    ANS: A   DIF: I   OBJ: 3-1.2

61. After air crosses a mountain, it sinks and
    a. cools, releasing moisture.
    b. cools, absorbing moisture.
    c. warms, releasing moisture.
    d. warms, absorbing moisture.

    ANS: D   DIF: I   OBJ: 3-1.2

62. People who live in coastal communities know that within a few blocks of the ocean the temperature can be cooler than further inland. This is because of
    a. altitude.
    b. surface currents.
    c. elevation.
    d. latitude.

    ANS: B   DIF: I   OBJ: 3-1.2

63. In which biome would you most likely find giraffes?
    a. temperate grasslands
    b. tropical savannas
    c. tundras
    d. tropical deserts

    ANS: B   DIF: I   OBJ: 3-2.2

64. In which climate zone is most of Africa located?
    a. tropical
    b. temperate
    c. polar
    d. antarctic

    ANS: A   DIF: II   OBJ: 3-2.1

65. In which climate zone is most of North America located?
    a. tropical
    b. temperate
    c. polar
    d. antarctic

    ANS: B   DIF: II   OBJ: 3-2.1

66. Which biome makes up most of Africa?
    a. tropical desert
    b. tropical rain forest
    c. chaparral
    d. tropical savanna

    ANS: D   DIF: II   OBJ: 3-2.2

67. Which biome makes up most of Canada?
    a. taiga
    b. tundra
    c. chaparral
    d. temperate forest

    ANS: A   DIF: II   OBJ: 3-2.2

68. The world's grasslands once covered about 42 percent of Earth's total land surface. Today, they occupy only about 12 percent. What percentage of Earth's grasslands do we have left?
    a. 12 percent
    b. 29 percent
    c. 42 percent
    d. 54 percent

    ANS: B   DIF: III   OBJ: 3-2.2

69. The Great Basin Desert is in a ____ of the Sierra Nevada.
    a. microclimate
    b. polar zone
    c. rain shadow
    d. tropical zone

    ANS: C   DIF: I   OBJ: 3-2.2

70. Plants are important to the atmosphere of Earth because they use ____ to make food.
    a. nitrogen
    b. carbon monoxide
    c. oxygen
    d. carbon dioxide

    ANS: D   DIF: I   OBJ: 3-3.3

COMPLETION

1. _____ is the condition of the atmosphere in a certain area over a long period of time. (Weather or Climate)

    ANS: Climate   DIF: I   OBJ: 3-1.1

2. _____ is the distance north and south from the equator measured in degrees. (Longitude or Latitude)

   ANS: Latitude DIF: I OBJ: 3-1.2

3. Savannas are grasslands located in the _____ zone between 23.5° north latitude and 23.5° south latitude. (temperate or tropical)

   ANS: tropical DIF: I OBJ: 3-2.1

4. Trees that lose their leaves are found in a(n) _____ forest. (deciduous or evergreen)

   ANS: deciduous DIF: I OBJ: 3-2.2

5. Frozen land in the polar zone is most often found in a _____. (taiga or tundra)

   ANS: tundra DIF: I OBJ: 3-2.2

6. A rise in global temperatures due to an increase in carbon dioxide is called _____. (global warming or the greenhouse effect)

   ANS: global warming DIF: I OBJ: 3-3.3

7. The day-to-day changes in temperature and precipitation define an area's _____. (weather or climate)

   ANS: weather DIF: I OBJ: 3-1.1

8. The equator and the lines running parallel to it are lines of _____. (latitude or elevation)

   ANS: latitude DIF: I OBJ: 3-1.2

9. Large grazing animals, including bison, live in the _____ grasslands of North America. (tropical or temperate)

   ANS: temperate DIF: I OBJ: 3-2.2

10. _____ have brightly colored leaves in autumn. (Deciduous trees or Evergreens)

    ANS: Deciduous trees DIF: I OBJ: 3-2.2

11. _____ blow mainly from one direction and affect the amount of precipitation region receives. (Surface currents or Prevailing winds)

    ANS: Prevailing winds DIF: I OBJ: 3-1.2

Holt Science and Technology
Copyright © by Holt, Rinehart and Winston. All rights reserved.

12. When warm air cools, it loses the ability to hold water vapor, which results in _____.

    ANS: precipitation   DIF: I   OBJ: 3-1.1

13. _____ is the height of surface landforms above sea level.

    ANS: Elevation   DIF: I   OBJ: 3-1.1

14. _____ currents are streamlike movements of water that occur at or near the surface of the ocean.

    ANS: Surface   DIF: I   OBJ: 3-1.2

15. A _____ is a large region characterized by a specific type of climate and the plants and animals that live there.

    ANS: biome   DIF: I   OBJ: 3-2.2

16. The _____ zone is the warm zone located around the equator.

    ANS: tropical   DIF: I   OBJ: 3-2.1

17. The _____ zone is the climate zone between the Tropics and the polar zone.

    ANS: temperate   DIF: I   OBJ: 3-2.1

18. The _____ zone includes the northernmost and southernmost climate zones.

    ANS: polar   DIF: I   OBJ: 3-2.1

19. _____ are trees that keep their leaves year-round.

    ANS: Evergreens   DIF: I   OBJ: 3-2.2

20. A permanently frozen layer of soil beneath thawed soil is called _____.

    ANS: permafrost   DIF: I   OBJ: 3-2.2

21. Evergreen needle-leaved trees, such as pine, spruce, and fir trees, are called _____.

    ANS: conifers   DIF: I   OBJ: 3-2.2

22. Small regions with unique climatic characteristics are called _____.

    ANS: microclimates   DIF: I   OBJ: 3-2.2

23. A period in which ice collects in high latitudes and moves toward lower latitudes is called a(n) _____.

    ANS: ice age   DIF: I   OBJ: 3-3.1

24. The period during an ice age in which enormous sheets of ice advance, getting bigger and covering a large area is called a _____ period.

    ANS: glacial   DIF: I   OBJ: 3-3.1

25. The period during an ice age in which ice begins to melt is called a(n) _____ period.

    ANS: interglacial   DIF: I   OBJ: 3-3.1

26. The _____ theory suggests that changes in the Earth's orbit and in the tilt of the Earth's axis cause ice ages.

    ANS: Milankovitch   DIF: I   OBJ: 3-3.2

27. The _____ is the Earth's natural heating process, in which gases in the atmosphere trap heat.

    ANS: greenhouse effect   DIF: I   OBJ: 3-3.3

28. _____ is the process of clearing forests.

    ANS: Deforestation   DIF: I   OBJ: 3-3.3

SHORT ANSWER

1. What is the difference between weather and climate?

   ANS:
   Weather is the condition of the atmosphere at a particular time and place. Climate is the average weather conditions in a certain area over a long period of time.

   DIF: I   OBJ: 3-1.1

2. How do mountains affect climate?

   ANS:
   Mountains can influence an area's temperature and precipitation. Temperature is affected by elevation. The windward side of a mountain receives much more precipitation than the rain-shadow side.

   DIF: I   OBJ: 3-1.1

3. Describe how air temperature is affected by ocean currents.

   ANS:
   The surface temperature of water affects the temperature of the air above it. As surface currents move, they carry warm or cool water to different locations. This changes the temperature of the air in the area.

   DIF: I   OBJ: 3-1.2

4. How would seasons be different if the Earth did not tilt on its axis?

   ANS:
   If the Earth did not tilt on its axis, there would be no seasons. The same amount of solar radiation would reach the Earth year-round.

   DIF: II   OBJ: 3-1.2

5. What are the soil characteristics of a tropical rain forest?

   ANS:
   The soil in a tropical rain forest is thin and nutrient poor. Nutrients are rapidly returned to the soil, but these nutrients are quickly absorbed and used by the plants. The remaining nutrients are washed away by heavy rains.

   DIF: I   OBJ: 3-2.2

6. In what way has savanna vegetation adapted to fire?

   ANS:
   Many plants require fire to reproduce. The seeds of some plants require fire to break open the seed's outer skin so the plant can grow. The heat from the fire triggers other plants to drop their seeds into the newly enriched soil.

   DIF: I   OBJ: 3-2.2

7. How do each of the tropical biomes differ?

   ANS:
   Answers will vary. Accept all reasonable responses. The tropical biomes differ in the amount of precipitation they receive and in their average temperature. This in turn affects the vegetation, the soil type, and the animals that live in each biome.

   DIF: I   OBJ: 3-2.2

8. Describe how tropical deserts and temperate deserts differ.

ANS:
Answers will vary. Sample answer: Tropical deserts are hot deserts, and temperate deserts are cold deserts. Winters in tropical deserts are usually mild, but temperate deserts often receive light snow during winter.

DIF: I   OBJ: 3-2.2

9. List and describe the three major climate zones.

ANS:
The three major climate zones are the tropical zone, the temperate zone, and the polar zone. The tropical climate zone receives the most direct solar radiation; therefore the temperatures are generally hot. Temperatures in the temperate zone tend to be moderate. The temperate zone experiences seasonal variations, such as warm summers and cold winters. During winter, the temperatures in the polar zone stay below freezing. During the summer, temperatures remain cold but they can be above freezing

DIF: I   OBJ: 3-2.1

10. Rank each biome according to how suitable it would be for growing crops. Explain your reasoning.

ANS:
Answers will vary.

DIF: II   OBJ: 3-2.2

11. a. How has the Earth's climate changed over time?
    b. What might have caused these changes?

ANS:
a. The Earth's climate has been both warmer and colder than it is today.
b. Climate changes may result from variations in the Earth's orbit; catastrophic events, such as volcanic eruptions; and the movement of the continents by plate tectonics and continental drift.

DIF: I   OBJ: 3-3.1

12. Explain how the greenhouse effect warms the Earth.

ANS:
Greenhouse gases allow sunlight to pass through the atmosphere, where it is absorbed by the Earth's surface and reradiated as thermal energy. The greenhouse gases absorb the thermal energy as it moves out of the atmosphere.

DIF: I   OBJ: 3-3.3

13. What are two ways that humans contribute to the increase in carbon dioxide levels in the atmosphere?

    ANS:
    burning fossil fuels and deforestation

    DIF: I   OBJ: 3-3.3

14. How will the warming of the Earth affect agriculture in different parts of the world?

    ANS:
    Answers will vary. The warming of the Earth would change the climate. Areas that were suitable for farming might become warmer and drier.

    DIF: I   OBJ: 3-3.3

15. Why are the poles colder than the equator?

    ANS:
    At the poles, the sun's rays strike the Earth's surface at a less direct angle than at the equator; the same amount of solar energy is spread over a larger area.

    DIF: I   OBJ: 3-1.2

16. Is precipitation more likely to occur when the prevailing winds are formed from warm air or when they are formed from cold air?

    ANS:
    Precipitation is more likely to occur when the prevailing winds are formed from warm air.

    DIF: I   OBJ: 3-1.2

17. Why are temperatures milder in Iceland than in Greenland?

    ANS:
    The Gulf Stream brings warm water to the ocean around Iceland. This warm water heats the air, making temperatures milder.

    DIF: I   OBJ: 3-1.2

18. Explain why the climate differs on opposite sides of a mountain range.

    ANS:
    When moving air hits a mountain range, it is forced upward. As the air rises, it cools, and much of the moisture it holds falls as precipitation of the windward side of the mountains. As the drier air descends on the other side of the mountains, it warms and draws up moisture from the surface, forming a desert in a rain shadow.

    DIF: I   OBJ: 3-1.2

Holt Science and Technology
Copyright © by Holt, Rinehart and Winston. All rights reserved.

19. Can a climate zone contain more than one biome?

    ANS:
    A climate zone may contain several different biomes.

    DIF: I   OBJ: 3-2.2

20. What is a microclimate?

    ANS:
    A microclimate is a small region with unique climate characteristics.

    DIF: I   OBJ: 3-2.2

21. Why does the sea level fall during glacial periods?

    ANS:
    The sea level falls during glacial periods because much of Earth's water is frozen during a glacial period.

    DIF: I   OBJ: 3-3.1

22. How might a major volcanic eruption have brought about an ice age?

    ANS:
    Dust, smoke, and ash from a volcanic eruption enter the atmosphere and act as a shield, blocking out much of the sun's rays and causing the Earth to cool.

    DIF: I   OBJ: 3-3.2

23. How might global warming affect coastal areas?

    ANS:
    The warmer temperatures could cause polar icecaps to melt, which would raise the sea level and cause flooding in coastal areas.

    DIF: I   OBJ: 3-3.3

24. Why do higher latitudes receive less solar radiation than lower latitudes?

    ANS:
    Higher latitudes receive less solar radiation because the sun's rays strike the Earth's surface at a less direct angle. This spreads the same amount of solar energy over a larger area, resulting in lower temperatures.

    DIF: I   OBJ: 3-1.2

Holt Science and Technology
Copyright © by Holt, Rinehart and Winston. All rights reserved.

25. How does wind influence precipitation patterns?

    ANS:
    The amount of precipitation an area receives can depend on whether the region's prevailing winds form from a warm air mass or from a cold air mass. If the winds form a warm air mass, they will probably carry moisture. If the winds form from a cold air mass, they will probably be dry. Precipitation is more likely to occur when the prevailing winds are warm and moist.

    DIF: I   OBJ: 3-1.2

26. Give an example of a microclimate. What causes the unique temperature and precipitation characteristics of this area?

    ANS:
    Answers will vary. Sample answer: Tundra and taiga biomes can be found on tropical mountains. This is because air at higher elevations retains less thermal energy than air at lower elevations.

    DIF: I   OBJ: 3-2.2

27. How have desert plants and animals adapted to this biome?

    ANS:
    Plants have adapted by developing fleshy leaves to store water and a waxy coating to prevent water loss. Animals are more active at night, when temperatures are cooler, and they burrow during the day.

    DIF: I   OBJ: 3-2.2

28. How are tundra and deserts similar?

    ANS:
    Both the tundra and desert biomes receive very little precipitation.

    DIF: I   OBJ: 3-2.2

29. Use the following terms to create a concept map: *climate, global warming, deforestation, greenhouse effect, flooding.*

    ANS:

    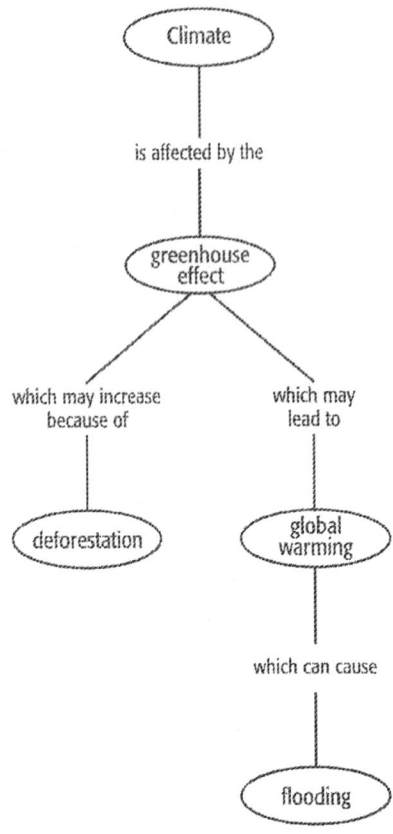

    DIF: II   OBJ: 3-3.3

30. Explain how ocean surface currents are responsible for milder climates.

    ANS:
    Warmer surface currents heat the surrounding air, and colder surface currents cool the surrounding air. A warm surface current might bring warmer temperatures to a cold area. A cold surface current might cool an area that is generally hot.

    DIF: II   OBJ: 3-1.2

31. In your own words, explain how a change in the Earth's orbit can affect the Earth's climates as proposed by Milutin Milankovitch.

    ANS:
    Answers will vary. Sample answer: A change in the Earth's orbit can affect the Earth's climates by limiting or increasing the amount of solar radiation the Earth receives.

    DIF: II   OBJ: 3-3.2

Holt Science and Technology
Copyright © by Holt, Rinehart and Winston. All rights reserved.

32. Explain why the climate differs drastically on each side of the Rocky Mountains.

    ANS:
    The climate differs on each side of the Rocky Mountains because the mountains affect the distribution of precipitation. The windward side receives more precipitation because as the warm air is forced to rise, it releases precipitation. As the dry air crosses the mountain, it sinks, warming and absorbing the moisture.

    DIF: II   OBJ: 3-1.2

33. What are some steps you and your family can take to reduce the amount of carbon dioxide that is released into the atmosphere?

    ANS:
    Answers will vary. Sample answer: We could ride bicycles when possible instead of using the car, and we could plant more trees.

    DIF: II   OBJ: 3-3.3

34. If the air temperature near the shore of a lake measures 24°C, and if the temperature increases by 0.05°C every 10 m traveled away from the lake, what would the air temperature be 1 km from the lake?

    ANS:
    1 km = 1,000 m   1000 m ÷ 10 m = 100
    100 × 0.05°C = 5°C   24°C + 5°C = 29°C

    DIF: II   OBJ: 3-1.2

The following illustration shows Earth's orbit around the sun.

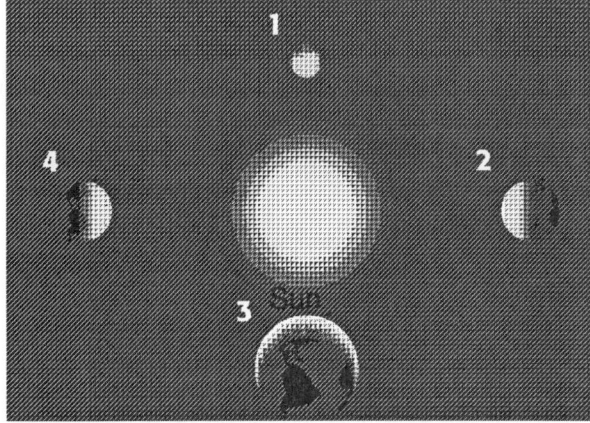

35. At which position **1**, **2**, **3**, or **4** is it spring in the Southern Hemisphere?

    ANS:
    3

    DIF: II   OBJ: 3-1.2

36. At which position does the South Pole receive almost 24 hours of daylight?

    ANS:
    2

    DIF: II   OBJ: 3-1.2

37. Explain what is happening in each climate zone in both the Northern Hemisphere and Southern Hemisphere at position 4.

    ANS:
    In the tropical zone temperatures are warm. The temperate zone in the Northern Hemisphere is experiencing summer. The temperate zone in the Southern Hemisphere is experiencing winter. Deciduous trees have shed their leaves. The polar zone in the Northern Hemisphere is experiencing almost 24 hours of daylight. Temperatures are cool, and the top meter of soil is thawing. The polar zone in the Southern Hemisphere is experiencing almost 24 hours of night. Temperatures are extremely cold, and the soil is still frozen.

    DIF: II   OBJ: 3-1.2

38. What are the three major climate zones?

    ANS:
    The three major climate zones are the tropical zone, the temperate zone, and the polar zone.

    DIF: I   OBJ: 3-2.1

39. The British Isles are approximately 2° farther north than Maine, yet Maine has a colder climate. Explain why the temperatures in the British Isles might be warmer than expected.

    ANS:
    Sample answer: The ocean current next to the British Isles is warm. The air coming off this warm water blows toward the islands, keeping their winter temperatures warmer than would be expected.

    DIF: II   OBJ: 3-1.2

40. Death Valley has an average annual temperature of about 24°C (76°F), but its climate is not pleasant. Explain how this average can be misleading.

    ANS:
    The average is misleading because summer days in this temperate desert are extremely hot, while nighttime and winter temperatures are low. The average does not represent the extreme hot and cold temperatures.

    DIF: II   OBJ: 3-2.2

41. The sun rose at 6:57 A.M. and set at 5:39 P.M. one day in a coastal city. How many hours and minutes of daylight did the city have that day? Show your work.

ANS:
Round 6:57 to 7:00 and 5:39 to 5:00. From 7:00 to 5:00 is 10 hours.
10 hours + 3 minutes + 39 minutes = 10 hours and 42 minutes

DIF: II   OBJ: 3-2.1

Below are climatograms of two different areas.

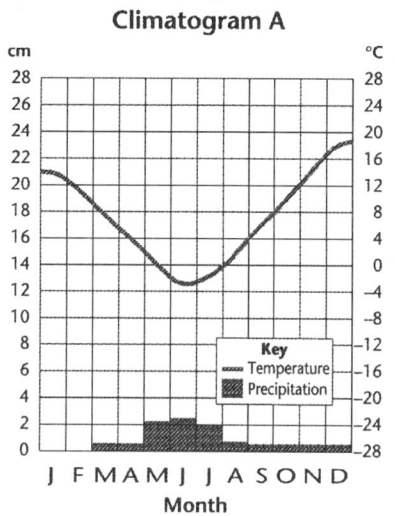
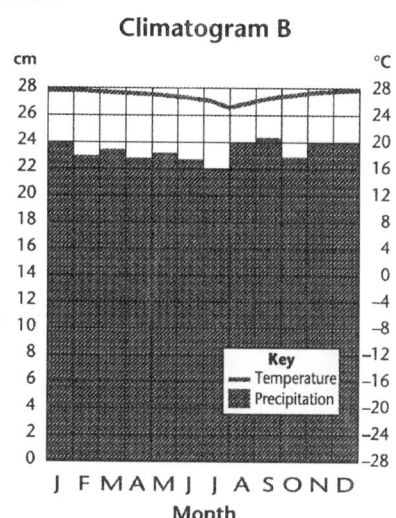

42. Identify the biome in which each climate can be found.

ANS:
Climatogram A is of a desert, and climatogram B is of a tropical rainforest.

DIF: II   OBJ: 3-2.2

43. How can you tell if the region represented in a climatogram is in the Northern Hemisphere, the Southern Hemisphere, or near the equator? Use the climatograms above as examples.

ANS:
If the temperatures are about the same year-round, as in climatogram B, then the area is near the equator. If the temperature is cool in June, July, and August, the area is in the Southern Hemisphere, as indicated by climatogram A. Warm temperatures in June, July, and August would indicate an area in the Northern Hemisphere.

DIF: II   OBJ: 3-1.2

44. Use the following terms to complete the concept map below: *carbon dioxide, polar regions, greenhouse gases, interglacial periods, burning fossil fuels, glacial periods, deforestation.*

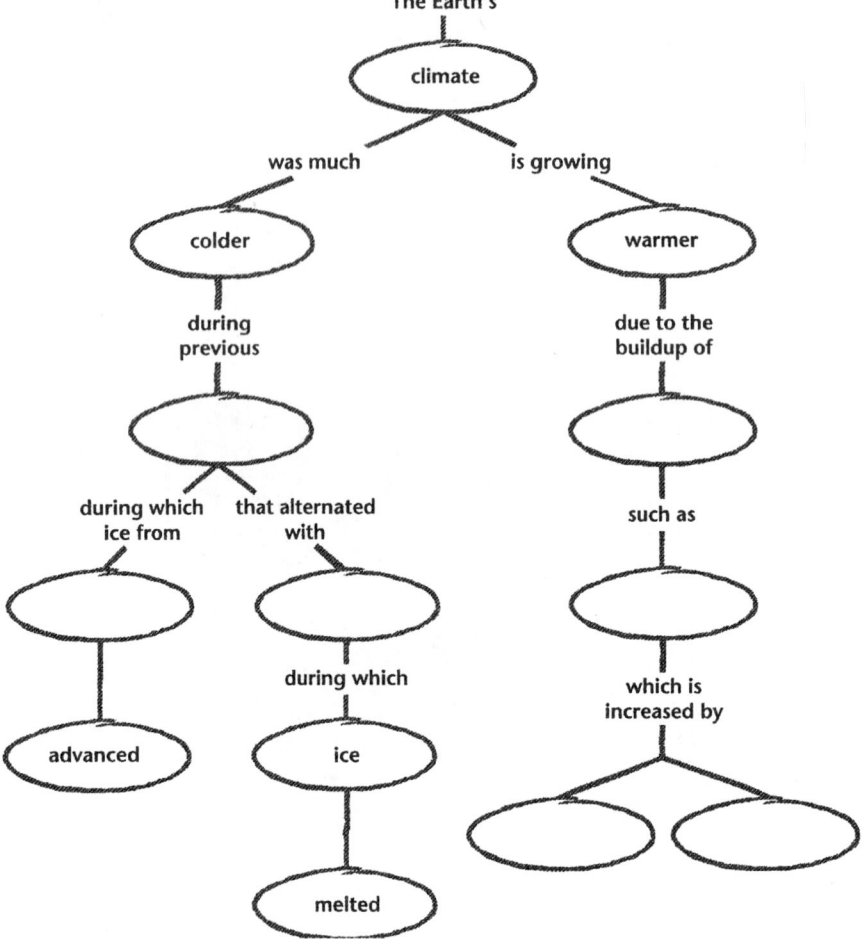

ANS:

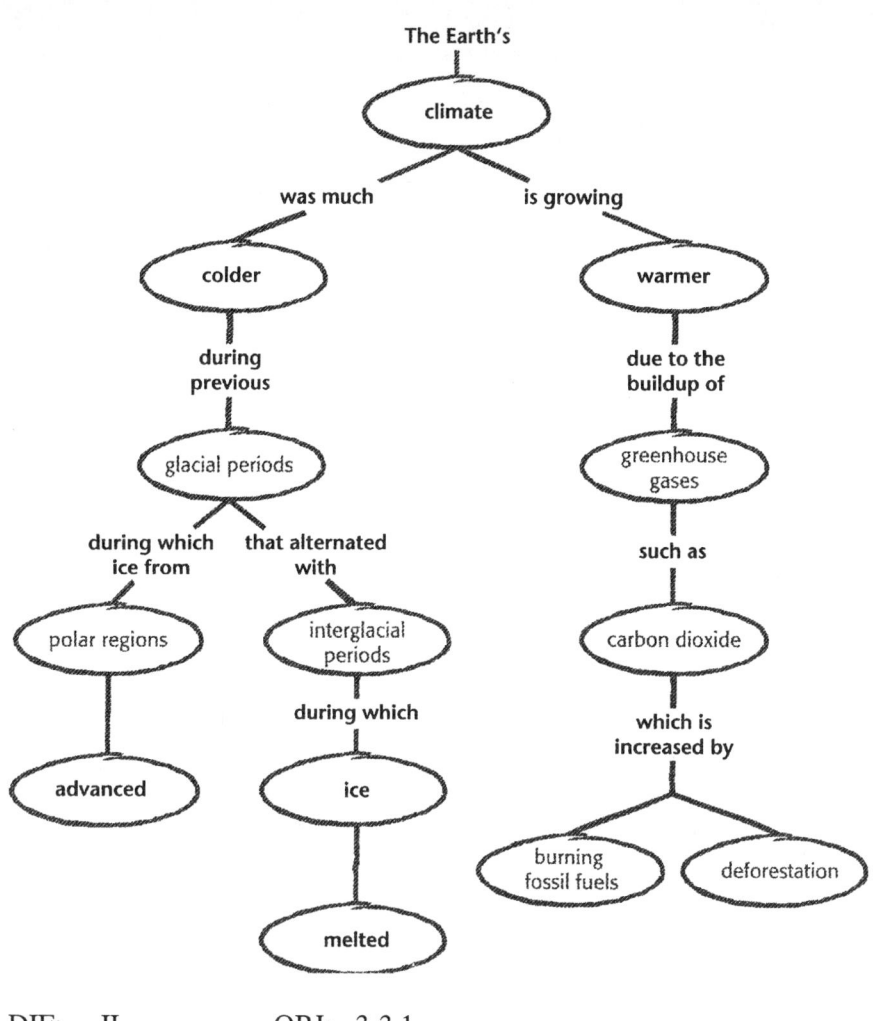

DIF: II OBJ: 3-3.1